About the cover: The cover displays the results of calculations in this book where there were measured values to compare with. There are also a number of calculations and predictions in this book where there exist no measured values to compare with, such as neutrinos and gravitons. Except for electrons, the calculated and predicted values are not exact because they are all dependent upon a measured value for $\alpha$ the Fine Structured Constant, which is not exact.

# *Advanced Electrino Physics*
## Draft 3

By Gordon L. Ziegler and Iris Irene Koch

# *Advanced Electrino Physics* Draft 3

ISBN-13: 978-1508638636
ISBN-10: 1508638632

This book was printed in the United States of America.

Rev. date: 02/27/2015

Authors
Gordon L. Ziegler and Iris Irene Koch
PO Box 1162
Olympia, WA 98507-1162 USA
e-mail: ben_ent100@msn.com

# PREFACE TO DRAFT 3

The principal reason for *Advanced Electrino Physics* Draft 3 is to correct errors in Draft 2 in the calculation of the masses of the pion family of particles. Of all the particles studied by the authors so far, the pion has been the hardest to get right.

Current models of physics, advanced as they are, cannot calculate the masses of elementary particles from first principles. The Electrino Fusion Model of Elementary Particles, however, is now at the stage when such calculations can be made. This book will calculate the g/2 factors of all the single particles used in particle structures up to state 4 or 5 in Chapter 5. These are necessary to calculate the masses of all the fundamental whole particles up to state 5. Except for 22 elementary particles calculated in this volume, the actual calculation of the masses of known particles shall be done in the set of three volumes *Predicting the Masses,* by the authors.

*Advanced Electrino Physics Draft 3* here also presents the theory of how to fuse the quartons in pions to the semions in electrons or positrons. This light is given to complete the knowledge base of the researcher. But this process is not pursued by the authors in that they think it is more hazardous than semion fusion, which could lead to many fatalities if it were tested.

An additional difference is discussed in this volume between total spin and observable spin. This intelligence should be useful to the researcher from now on in the calculations in particle physics.

In the previous volumes, *Electrino Physics* and *Advanced Electrino Physics,* different versions of CODATA database and Particle Data Group database were employed ranging from 1998 to 2006 dates. In Draft 2 and 3 of these volumes, the CODATA data is updated to the most recent 2010 version, and the Particle Data Group data is updated to the most recent 2013 and 2014 versions. The difference between the measured Fine Structure Constant $\alpha$ in the 2010 version and the 2006 version is in the right direction to harmonize the measured and calculated g/2 factors for the electron and muon, but it is not enough yet to harmonize them completely by this means. The inaccuracies of the data are because meters, kilograms, seconds, and Coulombs are not defined precisely enough. In natural units the data is already exact.

# CONTENTS

# Chapter 1

## FUSION OF QUARTONS

This book, *Advanced Electrino Physics* Draft 3, continues the presentation of the material started in the book, *Electrino Physics* Draft 2, but beyond the scope of that book. The first subject in that category is the fusion of quartons. While the author long believed that ionized quartons could fuse to semions, he did not know how to fuse bound quartons (as in pions, kaons, and D-ons), because those are zero spin particles and are bosons. They can go right through each other without colliding. Therefore they were not like positrons and electrons, of which the author theorized how to fuse the semions or anti-semions in them. The boson character of quarton systems presented a long insurmountable barrier in the author's mind to theorizing how to fuse quartons in bound systems.

The second crack in that barrier occurred as a result of conversations with Iris Koch, the author's sister, on the structure of pions. While ground state pions apparently have in them two quartons orbiting one way and two quartons orbiting the opposite way, we discussed a possible four-body state of the orbits being at right angles to each other, or other relative angles. We realized that pions, kaons, and D-ons could change relative orbit angles freely depending on energy state conditions.

The first crack in that barrier occurred years ago as a result of the author observing boson gravitons being ripped apart as the magnetic field in electrons tried to realign the orbits of the positrons and electrons in the gravitons, as seen in balancing decay schemes in chonomic equations for leptons.

On March 18, 2007, those ideas were put together in the author's mind. He first thought of realigning the quarton orbits in pions through a strong magnetic field. Seconds later he was impressed of doing it by means of a fast electron. He then looked in the 80th Edition of the *CRC Handbook of Chemistry and Physics*[1] to see if he could see any annihilation γγ decay modes for pions. He could see none. He could see none also for the D particle. But he found such a decay mode for kaons: $K^+ \to \pi^+\gamma\gamma$. He was at first surprised to see the $\pi^+$ particle in the decay products. But then he immediately saw that that $\pi^+$ did not come directly from the $K^+$ particle. It was the tag along left over particle with an electron in an unobserved electron neutrino colliding with the kaon. The electron was the active particle in the neutrino collision. First it realigned the orbits of the quartons in the kaon to make them fuse to two anti-semions in a positron. Second it annihilated with the resultant positron, leaving behind the left over pion. The pion in the reaction is a sure signature of the neutrino, proving this reaction did not occur naturally by other means.

Almost immediately the author thought of doing this artificially using high speed electrons rather than neutrinos. The electrons would be more controllable than neutrinos. In that case, $K^+ + e^- \to \gamma\gamma$. There would be no $\pi^+$ in this reaction. Using the skills taught in *Electrino Physics*[2] Chapter 10, Appendix A, and Appendix B, we can write chonomic equations for these two reactions:

Observed $K^+ \to \pi^+\gamma\gamma$.

```
               Collision,              annihi-
   K⁺   νₑ <0i  fusion    γ      γ      lation     π⁺      γ      γ
493..-i0.525eV q.p.1      0      0      q.p.2    139..     0      0
   |o        |       |    •|•    •|•    ••|••      |       •|•    •|•
   |_    +   |_   →   |_  + |_  +  |_  →   |_   →   |_  +   |_  +  |_
   |         |o       |o     |      |       |o       |o      |      |
   |        -|       -|+     |      |       |        |       |      |

 0|0      1|½     1|1   -1|-1  1|1   2-1|1     0|0     1|1   -1|-1
 0|0      ½|1     0|1    0|-1  0|1    0|1      1|1     0|1    0|-1
 |_____|       |_____|
   unobserved             unobserved
   particle               particles
```

An unobserved high speed electron neutrino, with $\tfrac{1}{2}\hbar$ positional-kinetic angular momentum, collides with a $K^+$. The magnetic field of the electron in the neutrino penetrates the $K^+$, which is made up of two orbiting pairs of quartons, and rotates the axes of the orbiting quartons to the same direction. The four quartons then fuse to two semions of a positron, which annihilates with the neutrino electron, leaving the neutrino $\pi^+$ and two oppositely directed energized pre-existing annihilaton photons (gammas).

# Theorized K⁺ + e⁻ → γγ.

```
              collision,                 annihi-
 K⁺      e⁻     fusion    γ      γ       lation     γ        γ
493...  0.51..   q.p.1    0      0       q.p.2      0        0
 _|o      _|      _|     •|•    •|•     ••|••      •|•      •|•
 _|  +   _|   →  _|   +  _|  +  _|  →    _|    →   _|   +   _|
  |      -|     -|+       |      |        |         |        |

 0|0     0|-½    0|0    -1|-1   1|1     1-1|0      1|1     -1|-1
 0|0     ½|0     0|0     0|-1   0|1      0|0       0|1      0|-1
                          |_____|
                           unobserved
                            particles
```

An axial spin, high speed electron, with ½ℏ positional-kinetic angular momentum, collides with a positive kaon. The magnetic field of the electron penetrates the kaon, which is made up of two orbiting pairs of quartons, and rotates the axes of the orbiting quartons to the same direction. The four quartons then fuse to two semions of a positron, which annihilates with the accelerated electron, leaving the oppositely directed energized pre-existing annihilation photons (gammas).

The first reaction is already observed and reported. This gives us confidence that the similar second reaction may take place as theorized. The reaction might be spin orientation sensitive. So the experimenter should not give up at a first failure. The end product is worth it.

By targeting anti-kaons with positrons, similarly to above, fusions and annihilations would occur. By targeting anti-kaons with electrons, stable electrons would be produced by anti-quarton fusion. No annihilation would occur. Also, in either case, the second law of thermodynamics would be reversed by these reactions, if the repetition rate were kept down. This could be an alternate method to that reported in Chapter 16 of *Electrino Physics*[3] for reversing the order to disorder arrow in the second law of thermodynamics. This might be done at an existing large accelerator laboratory. This may infuse new interest in the accelerators.

[1]SUMMARY TABLES OF PARTICLE PROPERTIES, January 1, 1998, Particle Data Group, as quoted by *CRC Handbook of Chemistry and Physics, 80th Edition*, David R. Lide, Editor-in Chief (Boca Raton: CRC Press, 1999), pp. **11**-1 to **11**-49.

[2]Gordon L. Ziegler, *Electrino Physics* (P.O. Box 1162, Olympia, WA 98507-1162 USA; e-mail: ben_ent100@msn.com: Book available for downloading free at http://benevolententerprises.org Book List. To order copies of this book, contact: Xlibris LLC, 1-888-795-4274, www.Xlibris.com, Orders@Xlibris.com. Draft 2 now available for purchase at CreateSpace.com and amazon.com and Kindle.)
[3]*Ibid.*

## Problem Set 1

1.      What characteristic of quarton systems presented a long insurmountable barrier in the author's mind to theorizing how to fuse quartons in bound systems?

2.      What is the second crack in that barrier that occurred?

3.      What is the first crack in that barrier that occurred years ago?

4.      How were those ideas put together in the author's mind?

5.      Did the author find an annihilation natural decay for pions?

6.      Why is that a tremendous blessing?

7.      What quarton particle did the author find that did have a listed annihilation natural decay?

8.      What unobserved particle is involved in the fusion of that particle.

9.      What additional unobserved particles are involved in the annihilation of that resultant particle?

10.     What value of positional-kinetic angular momentum occurs in the fusion of this particle that never occurred in any lepton decay in *Electrino Physics* Appendix A?

11.     What values of positional-kinetic angular momentum occurred in lepton decay in *Electrino Physics* Appendix A?

12.     In your opinion, is the value of positional-kinetic angular momentum in question 10 proper?  Is it possible?  Why?  Would any lower positive value of positional-kinetic angular momentum, other than zero, ever be observable?

13.     The $K^+ \rightarrow \pi^+ \gamma \gamma$ decay scheme occurs naturally.  In this reaction, does the $\pi^+$ come from the $K^+$ directly, as by knocking the $K^+$ echon down to the lower energy state?  Where does the $\pi^+$ in the above reaction come from?

14.     What decay scheme do we hope can be induced artificially?

15.     What would be the only observed products of such a reaction?

16.     Why, even if the reaction worked, might it not work the first time it was tried?

17.     How could stable electrons be produced by quarton fusion?

18.     Would such a reaction reverse the order to order disorder arrow in the second law of thermodynamics?

19.     Would this require a high repetition rate with a high beam current of anti-kaons, or a low repetition rate with a few anti-kaons?

20.    What would undoubtedly occur in the public perception of accelerators if this latter reaction occurred?

# Chapter 2

## HARMONIZING PARTICLE SPINS

In early editions of *Electrino Physics*, Chapter 6, in Postulate 8 and later derivations for the electron in Section IIIA, the calculated spin of the electron is $\hbar$, whereas the traditional spin of the electron is $\hbar/2$. The explanation in the model is that $\hbar$ is the total spin of the electron, whereas $\hbar/2$ is the observable spin of the electron.

The difference between total spin and observable spin is due to the structure of the particles. All particles are mass singularities. However much spin there may be inside a mass singularity, the only amount of spin that is observable from a mass singularity is due to the fracton on the side from which the spin is measured traveling at the speed of light at the event horizon of the mass singularity. The observer can observe no velocity faster than the speed of light, so he can observe only the spin that can be communicated at the event horizon of the singularity. The actual velocity of the fracton electrinos in the singularity may greatly exceed the speed of light (see the next chapter), but the greatest spin sense observable from the singularity can only be communicated at the event horizon of the black hole. Also, the observer cannot see *through* the mass singularity to see the fracton electrino on the opposite side of the singularity. That electrino contributes to the total spin of the particle, but not to the observable spin of the particle. For semion systems, the total spin may range from $\hbar$ to $\infty$ (see next chapter), but the observable spin is only always $\hbar/2$, because only one semion is observable at a time, with an effective mass of half the particle, at the radius r, and traveling at the maximum of the speed of light.

In the case of electrons, the two semions actually travel at the speed of light in their orbit. But their system is a mass singularity. The observer can only observe the effect of one semion. Half the mass of the electron times the radius of the electron times the speed of light c equals $\hbar/2$. That is the observable spin of the electron. The total spin of the electron takes

14

into consideration the contributions of both semions. We have, then, two times half the mass of the electron, or the mass of the electron times the radius of the electron times the speed of light c equals $\hbar$.

For a muon, the semions actually orbit in the mass singularity at 11.7062 c (see next chapter, Section I). This makes the total spin for the muon equal to $11.7062\,\hbar$. But the observable spin of the muon (and all simple semion systems) again, as explained above, is just $\hbar/2$. No matter what the energy state of the semion system, the observable spin is the same. This is the angular momentum that can be conveyed and transferred in particle collisions.

The spin dynamics of mass singularites are strange but simple. Mastering this bit of science will greatly help in the study of the rest of this book.

## Problem Set 2

1.    What is the total spin of the tauon?

2.    What is the observable spin of the tauon?

# Chapter 3

## PREDICTED MASSES OF CHARGED LEPTONS

So as this book may prepare the way for the calculation of the masses of every known particle, and may predict so far undetected particles, this book will here repeat, under a new title, Chapter 21 of *Electrino Physics.*

## A.     Introduction

In early chapters of *Electrino Physics,* the idea was expressed that electrons, muons, and tauons were just energy states of one particle system—and similarly for pions, Kaons, and D-ons as well as other particle sets. The author thought to solve for the various energy states like Niels Bohr solved for the energy states in hydrogen in 1913.[1] Bohr's calculational framework has been very helpful as a guide to the author in solving for the velocities, radii, and masses of particles in elevated states. This chapter will calculate these things. However there are many significant differences in the calculations. These will be pointed out.

## B.     The Bohr Atom

Bohr's results followed from algebraic derivations from a few postulates:

"1.     The electrons move in orbits restricted by the requirement that the angular momentum be an integral multiple of $h/2\pi$, that is, for circular orbits of radius r, the electron velocity v is restricted by

$$mvr = \frac{nh}{2\pi} \qquad (3\text{-}1)$$

and furthermore the electrons in these orbits do not radiate in spite of their acceleration. They were said to be in stationary states."[2]

"2.    Electrons can make discontinuous transitions from one allowed orbit to another, and the change in energy, E-E' will appear as radiation with frequency

$$\nu = \frac{E - E'}{h} \qquad (3\text{-}2)$$

An atom may absorb radiation by having its electrons make a transition to a higher energy orbit."[3]

3.    Bohr obtained another relevant calculational equation simply by balancing the Coulomb electric force against the centrifugal force.

$$\frac{kq_e^2}{r^2} = \frac{m_e v^2}{r}, \qquad (3\text{-}3)$$

where k = $1/(4\pi\varepsilon_0)$, and $q_e$ is the charge of the electron.[4]

4. "The energy of an electron in an orbit is the sum of its kinetic and potential energies:

$$E = E_{kinetic} + E_{potential} \qquad (3\text{-}4)$$

$$= \frac{1}{2}m_e v^2 - \frac{kq_e^2}{r}."[5] \qquad (3\text{-}5)$$

C.    Electron Energy Levels in Hydrogen

Performing simple algebraic operations, Bohr was able to solve for the orbital velocity v, the radius r, and the energy E.

"To begin, multiply both sides of Eq (3-3) by r to see

$$\frac{kq_e^2}{r} = m_e v^2. \qquad (3\text{-}6)$$

The term on the left hand side is the potential energy. So the equation for the energy becomes

$$E = \frac{1}{2} m_e v^2 - \frac{kq_e^2}{r} = -\frac{1}{2} m_e v^2. \qquad (3\text{-}7)$$

Now we just need to figure out what the velocity, v is equal to, so solve Eq (3-1) for r,

$$r = \frac{n\hbar}{m_e v}. \qquad (3\text{-}8)$$

Plug this into Eq (3-6),

$$kq_e^2 \frac{m_e v}{n\hbar} = m_e v^2. \qquad (3\text{-}9)$$

Then divide both sides by m_e v to see

$$\frac{kq_e^2}{n\hbar} = v. \qquad (3\text{-}10)$$

Now we can put in this value for v into the equation for energy, and then also plug in the values for k and ℏ, and we'll obtain the energy of the different levels of hydrogen:

$$E_n = \frac{-1}{2} m_e \left( \frac{kq_e^2}{n\hbar} \right)^2 \qquad (3\text{-}11)$$

$$= \frac{-1}{2} m_e \left( \frac{1}{4\pi\varepsilon_0} q_e^2 \frac{2\pi}{nh} \right)^2 \qquad (3\text{-}12)$$

18

$$= \frac{-m_e q_e^4}{8h^2 \varepsilon_0^2} \frac{1}{n^2}. \qquad (3\text{-}13)$$

Or, after substituting values for the constants,

$$E_n = \left(-13.6 \ eV\right)\frac{1}{n^2}. \qquad (3\text{-}14)$$

Thus, the lowest energy level of hydrogen (n = 1) is about -13.6 eV. The next energy level (n = 2) is -3.4 eV. The third (n = 3) is -1.51 eV, and so on. Note that these energies are less than zero, meaning that the electron is in a bound state with the proton. Positive energy states correspond to the ionized atom where the electron is no longer bound, but is in a scattering state."[6]


D.      Three More Quantum Numbers

"Bohr had pictured the electron orbits around the atomic center as being perfectly circular, but this was too simple. There are very few perfect circles in nature, and orbits in atoms are no exception.

"Later, in 1916, the German physicist Arnold Sommerfeld refined Bohr's 'easy' picture with one a bit more complex. In this modified view the electron orbits were not circular, but elliptical. But there are many kinds of ellipses possible (certainly more than one), and this changed the calculations in subtle ways, as each ellipse has a slightly different angular momentum. To take account of the possibility of elliptical orbits, Sommerfeld introduced another number; the **orbital quantum number** (sometimes called the "angular momentum quantum number"), which usually had the symbol "L" [or "l"]. . . .

"Like the principal quantum number, the orbital quantum number can have values of 0, 1, 2, 3, 4, etc., but only up to a whole number value of one **less** than the electron's principal quantum number (i.e. up to a value of n - 1). . . .

19

"There are two more quantum numbers associated with each electron; the **magnetic quantum number** written as **m**, and the **spin quantum number**, written as **s**. . . .

"To make it easy to picture what is going on, the magnetic quantum number can be thought of as defining the amount of "tilt" there is to the orbit.

"The possible values for **m** follow the same rules as for **L**, except that negative numbers are now allowed (the "tilt" of the orbit can be either "up" or "down"). So for **n = 2**, the possible values for **m** would be 0, 1, or -1. . . .

"There are only two possible values for **s** [spin] for any value of **n**. These values are usually written as +1/2 and -1/2, meaning either a clockwise spin or an anticlockwise spin.

"But what do these numbers tell us about the electrons?

"Austrian physicist Wolfgang Pauli worked out the significance of these numbers in 1925. He suggested that no two electrons in any given atom could have exactly the same values for all four quantum numbers.

"This became known as the **Pauli exclusion principle – 'No two electrons in any atom may have the same set of quantum numbers'.**"[7]


E.      Accuracies of the Models

The Neils Bohr Model was a close but not an exact fit to the measured data. The Sommerfeld Model of electron orbital ellipses, taking into account relativistic effects, gave a slightly better fit to the measured data. But as Thayer Watkins[8] demonstrates, no atomic structure model has a perfect fit to the measured data, and the Bohr Model is not much worse than the more advanced models. The fit is best between orbits of low quantum mechanical parameters n, and is worst between orbits of high quantum mechanical parameters n. At its best, the error can be as low as -0.01234 of 1%. But at its worst, the error can be at least as bad as -1.44546 of 1%.

These also are about the errors of the masses of charged leptons calculated in the next sections. Science has had 101 years to get the energies of the atom perfect, and has not done it. We should not hold back, therefore, until our model of energy states of semion orbits is perfect. We should publish a first cut mass model that is as close to the measured values as Bohr was to his measured values. The model we will advance in the next sections of the energies of semion orbits will be analogous to the Bohr model— not taking into account elliptical orbits, tilted orbits, or varying relativistic effects. There will be room for others to refine the model.

F.      Differences of Semion Orbits with Electron Orbits

There are a number of differences between semion orbits and electron orbits:
1. Semions have $e/2$ charge. Electrons have whole $e$ charge.
2. Each particle is a miniature black hole. The electron orbits as in Bohr's Model are exterior to black holes. The half charged particles called semions orbiting in charged leptons orbit inside black holes. This makes a difference in the force equation. The force for electrons in their orbit is $e^2/1 \cdot 4\pi\varepsilon_0\alpha^0 r^2$. The force on semions instead is $e^2/4 \cdot 4\pi\varepsilon_0\alpha^1 r^2$.
3. Semion orbits are a two body problem instead of a one body problem of electron orbits.
4. Semions orbit equal to or faster than the speed of light. Electrons orbit atoms slower than the speed of light.
5. The electron mass in the Bohr Model is the constant $m_e$. The semion mass in the outer non-relativistic frame is a variable.
6. Angular momentum in Bohr's atom is a function of n. But in our particles, the angular momentum is a function of not only n, but of $1/b$.
7. Between the models, standing waves have coincidence under different conditions. Instead of $C = n\lambda$ for electron orbits, $bC = n\lambda$ for semion orbits.
8. It is relatively easy to ionize electrons from orbit. It is virtually impossible to ionize semions from orbit. It is as

though the semions are contained in a speed of light boundary which they cannot pass.

9. Because the semion orbits, with the addition of the gravitational aether velocity v, are faster than the speed of light c, the electric force between the semions is reversed in sense, and the sign on the potential energy is changed.

G.      Deriving Particle States

Deriving particle states is one orbital level deeper than deriving electron orbital states that Niels Bohr did. The calculations are similar, but significantly different. Instead of treating the situation as a one body problem, we must treat the situation as a two body problem, with two equal semions in orbit about each other. This introduces an extra ½ into the expression of the centrifugal force.

We will solve for the particle states in a different order than Neils Bohr did. First we will balance the strong electric force in this problem with the central force of inertia in this problem, and familiarize ourselves with the whole problem to be solved:

Equation (3-16) below is the balancing of the force due to charge on the semions with the centrifugal force on the semions. The effective mass of a semion is half the mass of the whole particle in the outer non-relativistic frame. We use this mass of the semion in the centrifugal force along with the ½ from the two body problem. The velocity $v_o$ is greater than or equal to c, and must increase when the energy increases. In the electric type force side of the equation, the charge of the semion is e/2.

Different particle systems are in different order black holes. The force must depend on the order of black hole the particle system is in. Like the strong force and the electric force differ in strength by a power of $1/\alpha$, the forces in different orders of black holes differ by powers of $1/\alpha^{n/b}$. The electric type force expression, in the right side of Eq. (3-15), we expect to depend on a power of $1/\alpha^{n/b}$. To be in harmony with measured results and Eqn. (3-15), we want the power of $1/\alpha$ to be related to n/b. Also, we want the power for the electron to be such that the power of $\alpha$ is 1 when n = 0. We therefore take the power of $\alpha$ for electrons and

22

higher charged leptons to be n/b + 1. Completing the balancing of forces equation, we have

$$\frac{1}{2}\frac{m_e}{2}\frac{v_o^2}{r} = \left(\frac{e}{2}\right)^2\frac{1}{4\pi\varepsilon_0\alpha^{(n/b)+1}r^2}$$

(3-15)

The first ½ in the equation is from the two body nature of the problem, converting it to a one body problem. The $m_e/2$ is for the semion in ground state. The $v_o^2/r$ is the centrifugal force acting on the semions from a circular orbit. The e/2 is for the charge on each semion. The $4\pi\varepsilon_0$ are constants necessary to solve this problem in MKSC units. The $1/\alpha^{n/b}$ is the fine structure coupling constant between the orders of black holes in the problem. The r is the radius of the particle sub-particle orbit.

Thanks to the two body nature of the problem, all of the numeral constants except 4 in the above equation cancel out. $e^2/4\pi\varepsilon_0\alpha$ can be factored out as $\hbar c$. One r can cancel out of the two sides of the equation. The equation then looks like the following:

$$m_e v_o^2 = \frac{\hbar c}{\alpha^{n/b}r}$$

(3-16)

We can solve for $v_o$ if we can find an independent equation for r. Neils Bohr utilized the spin relation for electrons for this purpose. That equation was

$$m_e vr = n\hbar.$$

(3-17)

Unfortunately that spin relation does not work for charged leptons. Through trial and error, the author has settled on the following spin relation for charged leptons, which is here taken as a postulate:

$$m_e cr = n\hbar/b^2.$$

(3-18)

Solving for r we obtain:

$$r = n\hbar/b^2 m_e c. \qquad (3\text{-}19)$$

Combining Eqn. (3-19) with Eqn. (3-16) we have:

$$m_e v_0^2 = \hbar c b^2 m_e c / \alpha^{n/b} n\hbar \qquad (3\text{-}20)$$

$$v_0^2 = (b^2/n\alpha^{n/b})\, c^2 \qquad (3\text{-}21)$$

$$v_0 = (b^2/n\alpha^{n/b})^{1/2}\, c. \qquad (3\text{-}22)$$

The constant n in the above Eqn. is confusing. The n in Niels Bohr solution went 0, 1, 2, 3, 4, 5 . . . , but the n in the charged lepton solution goes 0, 1, 3, 6, 10, 15, 21 . . . , where b in the charged lepton solution goes 0, 1, 2, 3, 4, 5 . . . .

## H.    Deriving Semion Orbit Energy Levels and Masses

We have solved for $v_0$ in terms of our parameters n and b. We can now plug that formula into the relationship for particle energy to obtain the energy levels of semion orbits, and thus the particle masses. The kinetic, potential, and total energies of the semion system can be expressed as

$$Energy_{total} = Energy_{kinetic} + Energy_{potential}. \qquad (3\text{-}23)$$

Because, in some particles, like charges attract, the potential energy for charged leptons is positive instead of negative for orbiting electrons in Hydrogen.

$$Energy_{total} = 1/2\, m_e v_0^2 + m_e v_0^2 \qquad (3\text{-}24)$$

$$= 3/2\, m_e v_0^2 \qquad (3\text{-}25)$$

Substituting Eq. (3-22) into Eq. (3-25), we obtain

$$Energy_{total} = 3/2\, m_e (b^2/n\alpha^{n/b})\, c^2. \qquad (3\text{-}26)$$

The measurable mass term of the semion system is

24

$$M_T = 3/2\ (b_j^2/n_j\alpha_j^{n/b})\ m_e. \qquad (3\text{-}27)$$

The expression above in Eq. (3-27), derived from first principles, applies for a term in a series of terms for any charged lepton. But the mass of a particle equals that term plus a series of all previous terms back to that for the electron, where n and b equal zero (see Eq. (3-28)).

$$m_j =$$

$$\frac{3}{2}\left\{\left(\frac{b_j^2}{n_j\,\alpha^{n_j/b_j}}\right)g/2_j + \left(\frac{b_{j-1}^2}{n_{j-1}\,\alpha^{n_{j-1}/bj-1}}\right)g/2_{j-1} + ...1g_e/2\right\}m_e \qquad (3\text{-}28)$$

The right most term in Eqn. (3-28) without the g/2 factor for the electron is defined as 1.0.

To calculate this in general, we must have a definition of n, b, and j:

| j | 0 | 1 | 2 | 3 | 4 | 5 .... |
|---|---|---|---|---|---|---|
| n | 0 | 1 | 3 | 6 | 10 | 15 .... |
| b | 0 | 1 | 2 | 3 | 4 | 5 .... |

Table 3-1

The first three n and b are tested. Higher n and b are calculated. We expect both n and b to increase with j. We expect $n_j - (n_{j-1})$ to be $b_j$.

Finally, just as the mass is a series of terms, all other force terms are added by multiplying by half the g-factor for the given particle. For the muon, the net mass is the sum of the last two terms according to Eqn. (3-28) times half the g factor for the muon, or 206.553 998 611 59 $m_e$ times $|g_\mu/2|$, or 206.553 998 611 59 $m_e$ |- 1.001 165 912 4|, equals 206.794 822 47. . . . $m_e$, which is 1.000128, times the measured amount, 206.768 2843(52) $m_e$. That is 0.0128

25

of 1% error, which is almost the same error as the most accurate comparison of the theoretical Bohr Model of orbital differences to the measured values for orbital electrons. With what data we have to work with, our model is quite accurate.

There are only two usable measured g/2 factors that are available which can be used in calculating masses—for the electron and the muon. Fortunately, the match is close enough between particle masses and particle g/2 factors that calculated g/2 factors can be tested by the measured masses of the particles. We will begin employing calculated g/2 factors in this and the next two chapters.

I.      Fathoming the Orbital Velocities

Let us name the electron $e_0$, the muon $e_1$, and the tauon $e_2$. Then, for those particles and higher particles, Eq. (3-22) solves for the orbital semion velocities $v_o$:

| Particle | Semion Orbital Velocity $v_o$ |
|---|---|
| $e_0$ | 1.0000 c |
| $e_1$ | 11.7062 c |
| $e_2$ | 46.2480 c |
| $e_3$ | 167.8300 c |
| $e_4$ | 593.0600 c |
| $e_5$ | 2070.9000 c |

Table 3-2

Because the masses of the last three particles are so high as possibly to cause self fusion, the last three particles may not long exist. Compared to Einstein's Special Theory of Relativity, these are very high velocities. But these are velocities in a black hole. Velocities in a black hole should have no limit. Gravity escapes the bounds of a black hole, and communicates the sense of the mass of the black hole.

## J.      Theorizing the Radii of Semion Orbits

This model also theorizes the radii of the semion orbits in charged leptons:

Particle | Radius of Semion Orbit r

| | |
|---|---|
| $e_0$ | 1.0 ℏ/mc |
| $e_1$ | 1.0 ℏ/mc |
| $e_2$ | 3/4 ℏ/mc |
| $e_3$ | 6/9 ℏ/mc |
| $e_4$ | 10/16 ℏ/mc |
| $e_5$ | 15/25 ℏ/mc, |

where m is the mass of the given charged lepton—not just the mass of the electron.
Table 3-3

## K.      Predicted Masses of Charged Leptons

This model theorizes and predicts the masses of any charged leptons. Data are calculated from CODATA[9] 2010 α and particle masses from the same source. Predicted mass includes (mass term plus all previous predicted mass ratios to mass of the electron in MeV) times (particle g/2 factor).

| Charged Lepton | Mass Term (times $m_e$) | Predicted Mass | Measured Mass[9] |
|---|---|---|---|
| $e_0$ | 1.000 000 000 | 1.000 000 000 | 1.000 000 000 |
| $e_1$ | 205.553 998 611 | 206.794 822 479 | 206.768 2843(52) |
| $e_2$ | 3 208.351 934 | 3 420.101 469 | 3 477.15 |
| $e_3$ | 42 252.446 35 | 45 933.82738 | |
| $e_4$ | 527 591.655 3 | 577 827.418 5 | |

Table 3-4

This model allows that the masses of the last two particles may be so high as to cause self fusion, so they may not long exist.

The calculations in this chapter are for charged leptons. Similar calculations could be made for the quartons in the pion family. Other particle sets could be calculated by taking into account the Chonomic structures of the given particles. All the fundamental whole particles up to state 5 are calculated in the next chapter, from which all other particles up to state 5 may be calculated. Accompanying the calculation of fundamental masses in Chapter 4, is the calculation of the associated fundamental g/2 factors in Chapter 5.

---

[1]Stephen Gasiorowicz, *Quantum Physics* (New York: John Wiley & Sons, 1974), p. 15.

[2]*Ibid.*

[3]*Ibid.*

[4]"Bohr model," *Wikipedia*, the free encyclopedia, http://en.wikipedia.org/wiki/Bohr_model.

[5]*Ibid.*

[6]*Ibid.*

[7]Professor John Blamire, "Atomic Structure—The mystery of . . .—. . .the quantum atom," Exploring Life @ BIOdotEDU,

http://www.brooklyn.cuny.edu/bc/ahp/LAD/C3/C3_elecPos_02.html.

[8]Thayer Watkins, "The Relativistic Bohr Model of a Hydrogen-like Atom," applet-magic.com: Silicon Valley, Tornado Alley & BB Island USA, http://www.applet-magic.com/relabohr.htm.

9http://physics.nist.gov/cuu/Constants/.

## Problem Set 3

1.   How many quantum numbers are there in the Electrino Model of mass calculations of charged leptons?

2.   Does the author's model account for elliptical or tilted orbits?

3.   How accurate is the author's prediction of the mass of the muon?

4.   Why is the author's prediction of the mass of the tauon so inaccurate?

5.   What charge do semions have?

6.   What is the main difference between the force of electrons in their orbit and the force of semions in their orbit?

7.   How many body problem are semions in orbit?

8.   Do semions orbit slower than the speed of light or faster than the speed of light?

9.   Do charged lepton semion orbital velocities follow Einstein's Special Theory of Relativity?

10.    What difference in the mass is there between Bohr's Model and the author's model?

11.    What different rule for the coincidence of standing waves in the particle does the author's model have as compared to Bohr's Model?

12.    Are semions easy to ionize?

13.    What reverses the sense of the potential energy in semion orbits?

14.    What evidence is there that the mass in Eqn. (3-18) is the overall mass of the charged lepton, not just the mass of the electron?

15.    What forces are being balanced in Eqn. (3-19)?

16.    What factor occurs in the mass calculations due to both the kinetic and potential energies being positive?

17.    What mass term for the charged lepton is derived from first principles?

18.    How is that mass used in the calculation of the total mass of the particle?

19.    What is n as a function of j? What is b as a function of j?

20.    The Electrino Fusion Model of Elementary Particles predicts there will be how many different kinds of charged leptons in all?

21.    This is a free question: Did you imagine that $v_o$ should be so much above the speed of light for particles above electrons?

22.    This is a free question: Are the radii of semion orbits a function of $\alpha$? How?

# Chapter 4

## Redoing the Pion and Neutron

Originally Chapter 4 in previous drafts of this book was the simple re-publication of the journal article "Prediction of the Masses of Every Particle, Step 1." [1] However the first section of that article was the virtual repeat of the last chapter, and the last two sections (on the pion family and the neutron family) were incorrectly calculated. Therefore the authors replace that chapter with up to date calculations for the pion family and the neutron family.

## Redoing the Pion

### A. Pion Family Calculations

The author has had calculations for the pion family since practically the beginning of his work of calculating the masses of "elementary" particles or chonstructs. However, the pion has been the hardest to get right. The authors have published physics books with the wrong covers on them because of persistent confusion over the structure of the pion and the forces acting in it and on it. The authors published their revised structure of the pion, but now are returning to their original structure of the pion, which is and was two quartons orbiting one way and two quartons orbiting the opposite way, superimposed on the first pair of orbiting quartons. The two quartons in one pair of quartons approach the two quartons in the other quarton pair at a relative velocity of 2 c. 1) Do they collide with each other and reflect, initiating a continual chatter—do they sing? 2) Or, since their relative velocity is over c, do they attract each other to fusion to anti-semions in anti-electrons? 3) Or since the quartons have intrinsic 0 spin other than by their orbits, and their net orbital spin is 0, do they act as bosons and go through each other without colliding?

The pion appears to be stable. Thus option 2) must not occur for the pion, though it may occur for the excited state the kaon. By definition, the single quartons as well as their collective spins are bosons. In their lowest state they must act as stable bosons. Thus the quartons in their orbits must go through each other as if their opposing quartons were not even there. The cover of *Electrino Physics* Draft 2 shows the apparently correct structure of the pion among other particles.

According to *Predicting the Masses*, Volume 1, Introducing Chonomics, Chapter 9, postulate A.11. the mass of the particle term is the sort of exponential polynomial times the Energy Factor (EF) times the appropriate g/2 factor times the mass $m_e$. The exponential polynomial we can determine by solving for $v_0^2$ for the pion family member term. We first balance the strong electric force on one quarton with the inertial force of its circular orbit. The pion has two orbiting pairs of quartons. We first solve for one of those pairs.

The effective mass of a quarton is one fourth of the mass of the whole particle in the outer non-relativistic frame. We use this mass of the quarton in the centrifugal force along with the 1/2 from the two-body problem. The velocity $v_0$ is greater than $c$, and must increase when the energy increases. In the electric type force side of the equation, the charge of the quarton is $e/4$.

Each particle is a miniature mass singularity, and communicates with the outside world through powers of $\alpha$ (the Fine Structure Constant). The electric force expression, in the right side of Eqn. (4-1), we expect to depend on a power of $1/\alpha$. The numerator of the power of alpha n must be what makes the mass increase in the particle—namely the shells of mass from the radius $r_j$ to $r \to \infty$, which can be totaled by taking $(b+1)$ (pairing of shells) times $b/2$ (number of pairs of shells). The denominator in the power of $\alpha$ should be $b$ (the power of attenuation through $j$ orders of mass singularity). Also, we want the power for the quarton orbits to be such that the power of $\alpha$ is 1 when $n = (b^2 + b)/2 = 0$. We take the power of $\alpha$ for pions and higher charged pion family members to be $n/b+1$. To solve this in MKSC (SI) Units, we must put $4\pi\varepsilon_0$ in the denominator on the right side of the equation. Balancing the forces, we have

$$\frac{1}{2}\frac{m}{4}\frac{v_0^2}{r} = (e/4)^2 \Big/ 4\pi\varepsilon_0 \alpha^{(n/b)+1} r^2 \qquad (4\text{-}1)$$

for one orbiting quarton, where m is $m_e$. The $1/2$ in the equation is from the two-body nature of the problem, converting it to a one-body problem. The first $1/4$ in the equation is for the quarton non-relativistic effective mass. Some of the numeral constants in the above equation cancel out. The $e^2/4\pi\varepsilon_0\alpha$ can be factored out as $\hbar c$. One $r$ can cancel out of the two sides of the equation. The equation then looks like the following:

$$mv_0^2 = \hbar c / 2\alpha^{n/b} r, \text{ where } v_0 > c \quad \text{or} \quad n, b > 0, \qquad (4\text{-}2)$$

where m equals $m_e$. We can eliminate $m_e$ and r in this equation by combining this equation with Postulate A.6., *Predicting the Masses*, Volume 1, Introducing Chonomics, Chapter 9:

$$r = n^2\hbar/bm_ec \text{ (from postulate A.6. (Chapter 9)).} \qquad (4\text{-}3)$$

Combining Eqn. (4-3) with Eqn. (4-1), we can solve for $v_0^2$:

$$v_0^2 = (b/2n^2\alpha^{n/b})c^2, \quad b, n > 0, \qquad (4\text{-}4)$$

The $m_e$ and r cancel out in the above calculation. Both quarton orbits in pion family members have Eqn. (4-4) as the solution of the velocity squared for those orbits. But we are doing the quarton pairs one pair at a time. We will double the masses of the quarton pairs next. The sort of exponential polynomial in the parentheses is the first unit less term we need to calculate the mass terms of the pion family members. We will double this term in our mass evaluation table.

## B. The Kinetic, Potential, and Total Energies

The kinetic, potential, and total energies of the quarton system can be expressed as

$$\text{Energy}_{\text{total}} = \text{Energy}_{\text{kinetic}} + \text{Energy}_{\text{potential}} \qquad (4\text{-}5)$$

The potential energy of a mass term of a pion family member is

$$E_{\text{potential term}} = 2 \times (b/2n^2\alpha^{n/b})m_e c^2 \qquad (4\text{-}6)$$

Dividing the energy by $c^2$ we obtain the first term of a parameterized mass for the pion family member.

Mass is a volume thing, and is integrated from $r = r_j$ to $r \to \infty$ in discrete terms. The expression above in Eqn. (4-6), derived from first principles, is a term in a series of terms in a natural calculation of the mass of the pion family member. The sum of the terms is displayed in Table 4-1. See postulate A.12.

The Energy Factor (EF) for a single orbiting pair of quartons is 2 instead of 3/2. In electrons orbiting protons, the velocity v is much much less than c. In that case, the kinetic energy is ½ mv². But in the pion, the quartons orbit at c, and the kinetic energy is mc². That is a factor of two difference. The potential energy is also mc². The potential and kinetic energies are the same. So the EF factor in this case is 2.

The first author thought there should be no g/2-factors for the pion family members. But there must be $g/2$-factors after all for the pion family members. The strong term must be +1.0 instead of -1.0. And the most variable term is a term for the meso-electric force, not included in the electron $g/2$-factor. It should be

$$-bn\pi\alpha = -(p-1)n_{p-1}\pi\alpha. \qquad (14\text{-}7)$$

The g/2 factors for the pion family are predicted in *Advanced Electrino Physics* Draft 3, Chapter 5.

## C. Predictions of the Masses of the Quarton Pairs in the Pion Family

We now present Table 4-1 of the predicted (calculated from first principles) values of the masses of two quarton pairs in the pion family. In the Table, pions are denoted by $\pi_1$, kaons by $\pi_2$, D-ons by $\pi_3$, etc. Now we will double the mass calculations for the entire pion family member (see the exponential polynomial column in the table). Don't forget that the terms must be added to all previous terms to obtain the calculated predicted mass. Don't forget to multiply the result $m/m_e$ by the mass of the electron in MeV 0.510 998 928 to obtain the calculated mass of the particle (before the mass superposition calculation).

### Table 4-1
### (not including superposition calculation)

| Particle | b | n | $2 \times (b/2n^2\alpha^{n/b})$ | EF | g/2 factor | term $m/m_e$ | Calculated mass |
|---|---|---|---|---|---|---|---|
| Pion $\pi_1$ | 1 | 1 | 137.035 999 | 2 | 1.001 157 53 | 274.389 244 | 140.212 609 |
| Kaon $\pi_2$ | 2 | 3 | 356.483 548 | 2 | 0.978 240 60 | 697.453 360 | 496.610 528 |
| D-on $\pi_3$ | 3 | 6 | 1564.905 42 | 2 | 0.863 605 77 | 2702.922 70 | 1877.801 13 |

## D. Masses of the anti-pion family

The anti-pion family g/2 factors are the charge conjugates of the pion family g/2 factors, except they do not have terms for the meso-electric force. Otherwise, the mass tables for the anti-pion family are the same as the mass tables for the pion family. We will denote the anti-pion as $-\pi_1$.

### Table 4-2

| Particle | b | n | $2 \times (b/2n^2\alpha^{n/b})$ | EF | g/2 factor | Predicted $m/m_e$ | Calculated mass |
|---|---|---|---|---|---|---|---|
| $-\pi_1$ | 1 | 1 | 137.035 999 | 2 | -1.001 157 53 | -274.389 244 | -140.212 609 |
| $-\pi_2$ | 2 | 3 | 356.483 548 | 2 | -1.001 165 91 | -713.798 352 | -504.962 802 |
| $-\pi_3$ | 3 | 6 | 1564.905 42 | 2 | -1.001 157 65 | -3133.434 06 | -2106.144 24 |

35

These calculated masses are for both orbital pairs of quartons (see exponential polynomial term).

Now let us determine the active force between the oppositely spinning quarton pairs. Is it the strong nuclear force? The strong nuclear force is the force between nucleons like protons and neutrons, and pions are the mediating particle. Pions are strong gravitationally attracted entities to both nucleons, and are the go-between particles between the nucleons force wise. In nuclei, the pions have the leading important role. In the pion attached to a neutron to make it a proton, as in baryon structure in our chonomic system, is the strong nuclear force active with only one nucleon at a time (neutron or proton)? No. Pions mediate the strong nuclear force between nucleons, but a single pion is not attracted to a single neutron by the strong nuclear force.

What force then is active between the two halves of the pion? Is it the magnetic force? The magnetic force is active in this problem and adds a little to the maximum energy term a g/2 factor can have, but does not integrate well as a stick on force. According to the magnetic integration between two different levels of magnetism, the energy difference of the two levels calculates to be 0, because the magnetic field is a closed loop.

What force then is active between the two halves of the pion? The aether in the pion travels at or faster than c. So the strong electric force is active of like attracting like charges in the pion. The quartons in the pion are all positively charged. They attract each other. The integrable force in a pion is the virtual center of mass charge of one orbiting quarton pair attracting the other orbiting quartons. When the orbiting pairs coincide, the separation of the centers is not taken to be zero (which would lead to an infinite force), but r separation of one center to both the quartons in the opposite pair of quartons. The integral to be evaluated is

$$(2)\int_{\infty}^{r} \frac{e}{2} \frac{e}{4} \frac{1}{4\pi\varepsilon_0 \alpha r^2} \, dr = \frac{-\hbar c}{4r} + 0. \tag{4-12}$$

This is the integral of the strong electric force between the orbiting quarton pairs. In integrating a force times dr, we obtain the energy

36

difference between the quarton pairs from infinity separation to r separation. The (2) in the equation is for the two quartons in orbit 2 experiencing the force integration to the quarton pair 1 (we are taking both quartons together in this portion of the calculation to add to the double of the mass of the quarton orbits calculated previously); the e/2 is for the quarton pair 1 attracting quarton pair 2. The e/4 is for a single quarton charge attracted to quarton orbital pair 1 (doubled by (2)). We can combine Eqn. (4-3) with Eqn. (4-12) to eliminate the explicit r from the equation. We obtain

$$\Delta E = -bm_e c^2/4n^2 . \tag{4-13}$$

By dividing by $c^2$, we obtain the mass difference to add to the mass values in Tables 4-1 and 4-2.

$$\Delta m = -bm_e/4n^2. \tag{4-14}$$

We note that, except for the $m_e$, the above expression is similar to the sort of exponential polynomial for the orbits. The question comes, should there be also an EF factor and a g/2 factor in this radial integration? The expression in Eqn. (4-14) is for the potential energy difference between infinity and r. But a quarton accelerated from infinity to r would have a velocity also and a kinetic energy radially (not orbitally). The velocity would be negative. And times a positive velocity, the $v_r^2$ would be negative, and the kinetic energy would be the same as the potential energy. So it would be correct to expect that there should be an EF factor of 2 also for the radial integration and a g/2 factor.

## Table 4-3

| Pion family member | b | n | -b/4n² | EF | g/2 | integrated mass MeV | orbital mass MeV | total mass MeV | measured mass MeV [2] |
|---|---|---|---|---|---|---|---|---|---|
| Pion $\pi_1$ | 1 | 1 | -0.250 000 | 2 | 1.001 157 | -0.255 795 | 140.212 | 139.956 | 139.570 |
| Kaon $\pi_2$ | 2 | 3 | -0.055 555 | 2 | 0.978 240 | -0.055 542 | 496.610 | 496.554 | 493.677 |
| D-on $\pi_3$ | 3 | 6 | -0.020 833 | 2 | 0.863 605 | -0.018 387 | 1877.801 | 1877.78 | 1869.62 |

37

## Table 4-4

| Pion family member | b | n | -b/4n² | EF | g/2 | integrated mass MeV | orbital mass MeV | total mass MeV | measured mass MeV [2] |
|---|---|---|---|---|---|---|---|---|---|
| Anti-pion | 1 | 1 | -0.250 000 | 2 | -1.001 157 | 0.500 578 | -140.212 609 | -139.712 | not |
| Anti-kaon | 2 | 3 | -0.055 555 | 2 | -1.001 165 | 0.056 843 | -504.962 802 | -504.905 9 | measured |
| Anti-D-on | 3 | 6 | -0.020 833 | 2 | -1.001 157 | 0.021 316 | -2106.104 48 | -2106.083 | yet |

The calculated pion mass has three place accuracy to the measured pion mass. That is better than the original calculation. The kaon and D-on have two place accuracy compared with the measured masses.

---

### Redoing the Neutron

The third particle type—the neutron family—is different from the other two particle types. The neutron is a baryon, and, like all baryons, has affecting the mass calculations not only a relativistic imaginary-axis massive core, but also at the same time a real-axis zero mass core—the uniton. Electron family members orbit about this massive core particle. It is similar to electrons orbiting protons in Hydrogen. But electrons orbiting a proton are easily ionized, whereas the electron orbiting the uniton in a neutron is strongly bound to the core uniton. The uniton cannot come alone. For all lower particles, an electron always accompanies the uniton. Electrons orbiting protons could have semions in higher states, but this is not normally considered. Likewise the semions in the electron orbiting the uniton in neutrons could have higher states, but we do not consider them in our study of the neutron. It is the electron states orbiting the uniton in the neutron that can have a range of values in the electron orbits in the neutron.

The neutron has an interesting g/2 factor. Neutrons have positive cores, but they do not have positive strong force terms in their g/2 factors. They have negative strong force terms. [3](Chapter 5). What can account for this? Dots may be obscured by black holes, whereas – particles in the neutron (electrons), are

38

outside the black holes and not obscured. The +1 strong force of the dot is obscured. The -1 strong force of the − particle (electron orbiting the dot uniton) in the g/2 factor is not obscured. Therefore, to all appearances, the negative strong force replaces the positive strong force in the neutron. The same goes for the electric and magnetic terms.

The anti-neutron family has a + 1 strong force term from the orbiting positron (see above), no meso-electric term, and the charge conjugate of all the other terms as the neutron family member.

A binding orbit uses the g/2 factor of its central particle or the highest mass ratio sub-particle. Calculations are best if the energy factor and the g/2 factor are multiplied separately with each polynomial, and the mass ratios summed up.

Just as each previous particle family type has a different spin relation, the electron orbits of the uniton have a different spin relation. With the neutron family orbits, we will use B and N for the whole particle electron orbiting the uniton, to differentiate it from $b$ and $n$ in the electron family orbits (also used in these calculations). The total neutron spin is $mv_or$, but the observable spin on the event horizon of the black hole is only $mcr$. We have to go by the observable spin. The neutron observable spin relation is:

$$m_ecr = N\hbar/B \tag{4-15}$$

$$r = N\hbar/Bm_ec \tag{4-16}$$

Let us balance the force equation for the neutron family. Instead of the balancing of the electric force and the inertial force in Eqn. (4-17) starting with a 1/2, because of the two-body nature of such problems, Eqn. (4-17) starts with a '1', because of the single body nature of the neutron family problem. In this problem, the mass is m/1 instead of m/2, because in the main electron family orbits about the uniton, we are dealing with electrons as whole particles, not half particles.

$$1(m/1)v_o^2/r = (e/1)^2/4\pi\varepsilon_0\alpha^{(N/B)+1}r^2 \tag{4-17}$$

Using the techniques under Eqn. (4-1), this reduces to

$$mv_0^2 = \hbar c/\alpha^{N/B} r \quad . \tag{4-18}$$

Combining this with Eqn. (4-16), we obtain

$$mv_0^2 = \hbar c Bmc/N\hbar\alpha^{N/B} \tag{4-19}$$

$$v_{on}^2 = B/N\alpha^{N/B} c^2 \tag{4-20}$$

This is the orbital velocity squared for the neutron orbit of the electron family member. We have to combine that with the orbital velocity squared for the electron family member $v_{oe}^2$, which is solved by combining the electron spin relation substituting b for B and n for N in Eqn. (4-20), because the neutron space charge limits the spin relation for the intrinsic electron family members in the neutron, and is

$$v_{oe}^2 = b/n\alpha^{n/b} c^2 \tag{4-21}$$

There is a region where $v_{on}$ relative to $v_{oe}$ is faster, and a region where it is slower, but the average of the absolute value $v_{on}$ and average absolute value $v_{oe}$ are at right angles to each other and can be added by squaring them. The process is clarified in Eqn. (4-22).

$$v_{Tn}^2 = [(B/N\alpha^{N/B}) + (b/n\alpha^{n/b})] c^2 \tag{4-22}$$

By multiplying the quantity in Eq. (4-22) by $m_e$ we obtain the potential energy of the neutron family system. By multiplying the potential energy of the neutron family members by the EF $3/2$, we obtain the total energy including the kinetic energy in the neutron. By dividing by $c^2$ we obtain the mass $m$ of the neutron family member. Each particle has a $g/2$-factor. [1, 3] The result is in Eqn. (4-23).

$$m_{particle} = [(B_i/N\alpha^{N/B} \, EF \, g_i/2) + (b_j/n\alpha^{n/b} \, EF \, g_j/2)] \, m_e \tag{4-23}$$

Unitons are different from other whole-body systems. There are no elevated states of unitons. They are always only at state 2. The only elevated states associated with unitons are with orbiting particle systems surrounding the unitons. This feature of unitons apparently is reflected in the property that uniton systems have only one shell of mass. The sum of the velocity and thus mass components of the electron family member and the neutron system is not compounded by layers of mass shells. That typical stage of calculations will be left out of baryon calculations.

We see the mass of the neutron is a combination of two calculable terms. The N's and the n's are calculable from the B's and the b's. To calculate this in general, we must have a definition of n, b, and j: (See Table 4-5). The first three n and b are tested. Higher n and b are calculated. We expect both n and b to increase with j. We expect $n_j - (n_{j-1})$ to be $b_j$.

Table 4-5

| j | 0 | 1 | 2 | 3 | 4 | 5 | 6 |
|---|---|---|---|---|---|---|---|
| n | 0 | 1 | 3 | 6 | 10 | 15 | 21 |
| b | 0 | 1 | 2 | 3 | 4 | 5 | 6 |

For all neutron family members, the orbital spin is -1. The B for all neutron family members is 2. On the other hand, the minimum b and n are 0. The neutron family members all have $J = \frac{1}{2}$ and parity +. They all have unitons for core particles. They all have an electron family member with $\hbar/2$ intrinsic spin orbiting around the uniton with $\hbar$ orbital spin. The only things that differentiate the neutron family members are the energy states of the particles. Yet all the observed particles with these properties have different, seemingly unrelated, traditional names. Those observed so far are n, $\Lambda$, $\Sigma^0$, and $\Lambda_b^0$. To show the neutron related nature of those particles, we shall name those same particles $n = n_1$, $\Lambda = n_2$, $\Sigma^0 = n_3$, and $\Lambda_b^0 = n_4$, etc.

The g/2 terms are simplified also. There is only one g/2 factor for the B terms—at B = 2 ($\Lambda$ g/2 factor, see [3] Chapter 5). For the intrinsic states of the semion orbits in the orbiting electron about the uniton, see the electron family g/2 factors in [3] Chapter 5.

With the neutron family members, there is too much information to put in one table. We will divide the information into three tables.

| Particle | | b | n | $b/n\alpha^{n/b}$ | B | N | $B/N\alpha^{N/B}$ |
|---|---|---|---|---|---|---|---|
| n | $n_1$ | 0 | 0 | 1.000 000 000 | 2 | 3 | 1,069.450 645 |
| $\Lambda$ | $n_2$ | 1 | 1 | 137.035 999 7 | 2 | 3 | 1,069.450 645 |
| $\Sigma^0$ | $n_3$ | 2 | 3 | 1,069.450 645 | 2 | 3 | 1,069.450 645 |
| $\Lambda_b^0$ | $n_4$ | 3 | 6 | 9,389.432 523 | 2 | 3 | 1,069.450 645 |

Table 4-6  Neutron family parameters

In the next table, the g/2 factor to be utilized for B = 2 and N = 3 is the one where the meso-electric factor is -2 x $3\pi\alpha$, or the g/2 factor for the $\Lambda$ particle.

| Parti-cle | $b_i$ | $n_i$ | $g_i/2$ | $B_j$ | $N_j$ | $g_j/2$ |
|---|---|---|---|---|---|---|
| n | 0 | 0 | -1.000 000 000 | 2 | 3 | -1.138 716 794 |
| $\Lambda$ | 1 | 1 | -1.001 165 912 | 2 | 3 | -1.138 716 794 |
| $\Sigma^0$ | 2 | 3 | -1.001 157 653 | 2 | 3 | -1.138 716 794 |
| $\Lambda_b^0$ | 3 | 6 | -1.001 165 744 | 2 | 3 | -1.138 716 794 |

Table 4-7  System g/2 factors

| Particle | Predicted Mass Ratio (m/m_e) | Measured Mass Ratio [2] (m/m_e) |
|---|---|---|
| n | 1,828.202 115 | 1,838.683 66 |
| $\Lambda$ | 2,032.495 772 | 2,183.337 |
| $\Sigma^0$ | 2,288.490 108 | 2,333.9 |
| $\Lambda_b^0$ | 10,618.179 61 | 10,996 |

Table 4-8

All calculated values, except for the Λ, are two place accuracy to the measured values. The calculated mass ratios of each neutron family member are a little on the low side compared to the measured values. This is to be expected because we are calculating only circular orbits and neglecting relativistic effects. But actually, our calculated values are a pretty good fit with the measured values. When our calculated values go up by small steps, the measured values go up by small steps. And when our calculated value goes up by a large step, the measured value goes up by a large step—with about the same degree of precision.

---

[1]    Gordon L. Ziegler and Iris Irene Koch, "Prediction of the Masses of Every Particle, Step 1," **Galilean Electrodynamics**, Summer 2010, Vol. 21, SI No.3, pp. 43-49.

[2]    J. Beringer **et al.** (Particle Data Group) PR **D86**, 010001 (2012) and 2013 update for the 2014 edition (URL: http://pdg.lbl.gov).

[3]    Gordon L. Ziegler and Iris Irene Koch, *Advanced Electrino Physics* Draft 2. Draft 3 now available at Amazon.com.

# Chapter 5

## PATTERN g/2 FACTORS

This chapter was calculated before the advanced light in Chapter 4 on the method of the calculation of the neutron family particle mass ratios was determined. As it turned out, we only needed one g/2 factor for the electrons orbiting the uniton. But instead of gutting the g/2 factors in in this chapter for the neutron and anti-neutron families, to match the Chapter 4 results, we preserve all the original g/2 factor calculations in this chapter in case we need them in calculating subsequent multi-particle particles, for the g/2 factors are very difficult to calculate, and we do not want to have to calculate them again.

This chapter succeeds "An Update on g/2 Factors" in the original Chapter 5 in *Advanced Electrino Physics*. That paper was a timely advance in the field, but it contained errors. Also it was incomplete. It had g/2 factors for fundamental matter particles— the electron family, the pion family, and the neutron family. But it did not include g/2 factors for their anti-particles, which, because of the meso-electric terms for positive but not negative particles, are not simple charge conjugates to the matter particles. This chapter corrects the original matter g/2 factors in Chapter 5, and adds the corresponding g/2 factors for the antimatter fundamental particles. With all the g/2 factors in this chapter, you can calculate any particle mass up to state 4 or 5 in the electron, pion, and neutron families and their anti-particles.

Table 5-1
Electron family
Electron $e_0$ g/2 Factor Evaluation with 2010 $\alpha$

| force: | g/2 factor term: | numerical value: |
|---|---|---|
| strong | -1 | -1.000 000 000 000 |
| electric | $-\alpha/2\pi$ | -0.001 161 409 733 |
| magnetic | $-\alpha^2/16\pi^2$ | -0.000 000 337 218 |
| $weak_1$ | $+\alpha^2/8\pi$ | +0.000 002 118 804 |
| $weak_2$ | $-\alpha^3/4\pi$ | -0.000 000 030 923 |
| $weak_3$ | $+(32\alpha)^1\,\alpha^3/4\pi$ | +0.000 000 007 221 |
| $weak_4$ | $-(32\alpha)^3\,\alpha^3/4\pi$ | -0.000 000 000 393 |
| $weak_5$ | $+(32\alpha)^4\,\alpha^3/4\pi$ | +0.000 000 000 091 |
| $weak_6$ | $-(32\alpha)^5\,\alpha^3/4\pi$ | -0.000 000 000 021 |
| $weak_7$ | $+(32\alpha)^6\,\alpha^3/4\pi$ | +0.000 000 000 005 |
| $weak_8$ | $-(32\alpha)^7\,\alpha^3/4\pi$ | -0.000 000 000 001 |
| $weak_9$ | $+(32\alpha)^8\,\alpha^3/4\pi$ | +0.000 000 000 000 |
| total calculated $g_e/2$ | | -1.001 159 652 169 |
| compare measured $g_e/2$ | | -1.001 159 652 181 1(08) [1] |

Table 5-2
Electron family
Muon $e_1$ g/2 Factor Evaluation with 2010 $\alpha$

| force: | g/2 factor term: | numerical value: |
|---|---|---|
| strong | $-1$ | -1.000 000 000 000 |
| electric | $-\alpha/2\pi$ | -0.001 161 409 733 |
| magnetic | $-\alpha^2/16\pi^2$ | -0.000 000 337 218 |
| $weak_1$ | $-\alpha^2/4\pi$ | -0.000 004 237 608 |
| $weak_2$ | $+3\alpha^3/4\pi$ | +0.000 000 092 769 |
| $weak_3$ | $-3(32\alpha)^1 \alpha^3/4\pi$ | -0.000 000 021 663 |
| $weak_4$ | $+3(32\alpha)^3 \alpha^3/4\pi$ | +0.000 000 001 181 |
| $weak_5$ | $-3(32\alpha)^4 \alpha^3/4\pi$ | -0.000 000 000 275 |
| $weak_6$ | $+3(32\alpha)^5 \alpha^3/4\pi$ | +0.000 000 000 064 |
| $weak_7$ | $-3(32\alpha)^6 \alpha^3/4\pi$ | -0.000 000 000 015 |
| $weak_8$ | $+3(32\alpha)^7 \alpha^3/4\pi$ | +0.000 000 000 003 |
| $weak_9$ | $-3(32\alpha)^8 \alpha^3/4\pi$ | -0.000 000 000 000 |
| $weak_{10}$ | $+3(32\alpha)^9 \alpha^3/4\pi$ | +0.000 000 000 000 |

total calculated g/2 factor for muon  -1.001 165 912 495
compare measured g/2 factor muon  -1.001 165 920 7(06)

Table 5-3
Electron family
Tauon $e_2$ g/2 Factor Evaluation with 2010 $\alpha$

| force: | g/2 factor term: | numerical value: |
|---|---|---|
| strong | $-1$ | -1.000 000 000 000 |
| electric | $-\alpha/2\pi$ | -0.001 161 409 733 |
| magnetic | $-\alpha^2/16\pi^2$ | -0.000 000 337 218 |
| $weak_1$ | $+\alpha^2/4\pi$ | +0.000 004 237 608 |
| $weak_2$ | $-6\alpha^3/4\pi$ | -0.000 000 185 539 |
| $weak_3$ | $+6(32\alpha)^1\alpha^3/4\pi$ | +0.000 000 043 663 |
| $weak_4$ | $-6(32\alpha)^3\alpha^3/4\pi$ | -0.000 000 002 362 |
| $weak_5$ | $+6(32\alpha)^4\alpha^3/4\pi$ | +0.000 000 000 551 |
| $weak_6$ | $-6(32\alpha)^5\alpha^3/4\pi$ | -0.000 000 000 128 |
| $weak_7$ | $+6(32\alpha)^6\alpha^3/4\pi$ | +0.000 000 000 030 |
| $weak_8$ | $-6(32\alpha)^7\alpha^3/4\pi$ | -0.000 000 000 007 |
| $weak_9$ | $+6(32\alpha)^8\alpha^3/4\pi$ | +0.000 000 000 001 |
| $weak_{10}$ | $-6(32\alpha)^7\alpha^3/4\pi$ | -0.000 000 000 000 |

total calculated g/2 factor for tauon    -1.001 157 653 136

Table 5-4
Electron family
$e_3$ g/2 Factor Evaluation with 2010 $\alpha$

| force: | g/2 factor term: | numerical value: |
|---|---|---|
| strong | $-1$ | -1.000 000 000 000 |
| electric | $-\alpha/2\pi$ | -0.001 161 409 733 |
| magnetic | $-\alpha^2/16\pi^2$ | -0.000 000 337 218 |
| $weak_1$ | $-\alpha^2/4\pi$ | -0.000 004 237 608 |
| $weak_2$ | $+10\alpha^3/4\pi$ | +0.000 000 309 233 |
| $weak_3$ | $-10(32\alpha)^1\,\alpha^3/4\pi$ | -0.000 000 072 210 |
| $weak_4$ | $+10(32\alpha)^3\,\alpha^3/4\pi$ | +0.000 000 005 937 |
| $weak_5$ | $-10(32\alpha)^4\,\alpha^3/4\pi$ | -0.000 000 001 919 |
| $weak_6$ | $+10(32\alpha)^5\,\alpha^3/4\pi$ | +0.000 000 000 214 |
| $weak_7$ | $-10(32\alpha)^6\,\alpha^3/4\pi$ | -0.000 000 000 050 |
| $weak_8$ | $+10(32\alpha)^7\,\alpha^3/4\pi$ | +0.000 000 000 011 |
| $weak_9$ | $-10(32\alpha)^8\,\alpha^3/4\pi$ | -0.000 000 000 002 |
| $weak_{10}$ | $+10(32\alpha)^9\,\alpha^3/4\pi$ | +0.000 000 000 000 |
| $weak_{11}$ | $-10(32\alpha)^{10}\,\alpha^3/4\pi$ | -0.000 000 000 000 |

total calculated g/2 factor for $e_3$     -1.001 165 744 345

Table 5-5
Electron family
$e_4$ g/2 Factor Evaluation with 2010 α

| force: | g/2 factor term: | numerical value: |
|---|---|---|
| strong | $-1$ | $-1.000\ 000\ 000\ 000$ |
| electric | $-\alpha/2\pi$ | $-0.001\ 161\ 409\ 733$ |
| magnetic | $-\alpha^2/16\pi^2$ | $-0.000\ 000\ 337\ 218$ |
| $weak_1$ | $+\alpha^2/4\pi$ | $+0.000\ 004\ 237\ 608$ |
| $weak_2$ | $-15\alpha^3/4\pi$ | $-0.000\ 000\ 463\ 849$ |
| $weak_3$ | $+15(32\alpha)^1\,\alpha^3/4\pi$ | $+0.000\ 000\ 103\ 316$ |
| $weak_4$ | $-15(32\alpha)^3\,\alpha^3/4\pi$ | $-0.000\ 000\ 005\ 633$ |
| $weak_5$ | $+15(32\alpha)^4\,\alpha^3/4\pi$ | $+0.000\ 000\ 001\ 315$ |
| $weak_6$ | $-15(32\alpha)^5\,\alpha^3/4\pi$ | $-0.000\ 000\ 000\ 307$ |
| $weak_7$ | $+15(32\alpha)^6\,\alpha^3/4\pi$ | $+0.000\ 000\ 000\ 071$ |
| $weak_8$ | $-15(32\alpha)^7\,\alpha^3/4\pi$ | $-0.000\ 000\ 000\ 016$ |
| $weak_9$ | $+15(32\alpha)^8\,\alpha^3/4\pi$ | $+0.000\ 000\ 000\ 003$ |
| $weak_{10}$ | $-15(32\alpha)^9\,\alpha^3/4\pi$ | $-0.000\ 000\ 000\ 000$ |
| $weak_{11}$ | $+15(32\alpha)^{10}\,\alpha^3/4\pi$ | $+0.000\ 000\ 000\ 000$ |

total calculated g/2 factor for $e_4$    $-1.001\ 167\ 869\ 444$

Table 5-6
Positron family
Anti-electron -$e_0$ g/2 Factor Evaluation with 2010 $\alpha$

| force: | g/2 factor term: | numerical value: |
|---|---|---|
| strong | $+1$ | $+1.000\ 000\ 000\ 000$ |
| meso-electric | $-bn\pi\alpha$ | $-0.000\ 000\ 000\ 000$ |
| electric | $+\alpha/2\pi$ | $+0.001\ 161\ 409\ 733$ |
| magnetic | $+\alpha^2/16\pi^2$ | $+0.000\ 000\ 337\ 218$ |
| weak$_1$ | $-\alpha^2/8\pi$ | $-0.000\ 002\ 118\ 804$ |
| weak$_2$ | $+\alpha^3/4\pi$ | $+0.000\ 000\ 030\ 923$ |
| weak$_3$ | $-(32\alpha)^1\,\alpha^3/4\pi$ | $-0.000\ 000\ 007\ 221$ |
| weak$_4$ | $+(32\alpha)^3\,\alpha^3/4\pi$ | $+0.000\ 000\ 000\ 393$ |
| weak$_5$ | $-(32\alpha)^4\,\alpha^3/4\pi$ | $-0.000\ 000\ 000\ 091$ |
| weak$_6$ | $+(32\alpha)^5\,\alpha^3/4\pi$ | $+0.000\ 000\ 000\ 021$ |
| weak$_7$ | $-(32\alpha)^6\,\alpha^3/4\pi$ | $-0.000\ 000\ 000\ 005$ |
| weak$_8$ | $+(32\alpha)^7\,\alpha^3/4\pi$ | $+0.000\ 000\ 000\ 001$ |
| weak$_9$ | $-(32\alpha)^8\,\alpha^3/4\pi$ | $-0.000\ 000\ 000\ 000$ |

total calculated g/2 factor for $-e_0$     $+1.001\ 159\ 652\ 169$

Table 5-7
Positon family
Anti-muon -e$_1$ g/2 Factor Evaluation with 2010 α

| force: | g/2 factor term: | numerical value: |
|---|---|---|
| strong | $+1$ | +1.000 000 000 00 |
| meso-electric | $-bn\pi\alpha$ | -0.022 925 309 22 |
| electric | $+\alpha/2\pi$ | +0.001 161 409 73 |
| magnetic | $+\alpha^2/16\pi^2$ | +0.000 000 337 21 |
| weak$_1$ | $+\alpha^2/4\pi$ | +0.000 004 237 60 |
| weak$_2$ | $-3\alpha^3/4\pi$ | -0.000 000 092 76 |
| weak$_3$ | $+3(32\alpha)^1\alpha^3/4\pi$ | +0.000 000 021 66 |
| weak$_4$ | $-3(32\alpha)^3\alpha^3/4\pi$ | -0.000 000 001 18 |
| weak$_5$ | $+3(32\alpha)^4\alpha^3/4\pi$ | +0.000 000 000 27 |
| weak$_6$ | $-3(32\alpha)^5\alpha^3/4\pi$ | -0.000 000 000 06 |
| weak$_7$ | $+3(32\alpha)^6\alpha^3/4\pi$ | +0.000 000 000 01 |
| weak$_8$ | $-3(32\alpha)^7\alpha^3/4\pi$ | -0.000 000 000 00 |
| weak$_9$ | $+3(32\alpha)^8\alpha^3/4\pi$ | +0.000 000 000 00 |
| weak$_{10}$ | $-3(32\alpha)^9\alpha^3/4\pi$ | -0.000 000 000 00 |

total calculated g/2 factor for $-e_1$     +0.978 240 603 27


* b is advanced 1.

Table 5-8
Positron family
Anti-tauon -$e_2$ g/2 Factor Evaluation with 2010 $\alpha$

| force: | g/2 factor term: | numerical value: |
|---|---|---|
| strong | $+1$ | +1.000 000 000 0 |
| meso-electric | $-bn\pi\alpha$ | -0.137 551 855 3 |
| electric | $+\alpha/2\pi$ | +0.001 161 409 7 |
| magnetic | $+\alpha^2/16\pi^2$ | +0.000 000 337 2 |
| $weak_1$ | $-\alpha^2/4\pi$ | -0.000 004 237 6 |
| $weak_2$ | $+6\alpha^3/4\pi$ | +0.000 000 185 5 |
| $weak_3$ | $-6(32\alpha)^1\alpha^3/4\pi$ | -0.000 000 043 6 |
| $weak_4$ | $+6(32\alpha)^3\alpha^3/4\pi$ | +0.000 000 002 3 |
| $weak_5$ | $-6(32\alpha)^4\alpha^3/4\pi$ | -0.000 000 000 5 |
| $weak_6$ | $+6(32\alpha)^5\alpha^3/4\pi$ | +0.000 000 000 1 |
| $weak_7$ | $-6(32\alpha)^6\alpha^3/4\pi$ | -0.000 000 000 0 |
| $weak_8$ | $+6(32\alpha)^7\alpha^3/4\pi$ | +0.000 000 000 0 |
| $weak_9$ | $-6(32\alpha)^8\alpha^3/4\pi$ | -0.000 000 000 0 |
| $weak_{10}$ | $+6(32\alpha)^9\alpha^3/4\pi$ | +0.000 000 000 0 |

total calculated g/2 factor for $-e_2$     +0.863 605 799 0

Table 5-9
Positron family
-e$_3$ g/2 Factor Evaluation with 2010 α

| force: | g/2 factor term: | numerical value: |
|---|---|---|
| strong | $+1$ | $+1.000\ 000\ 000\ 000$ |
| meso-electric | $-bn\pi\alpha$ | $-0.412\ 655\ 679\ 116$ |
| electric | $+\alpha/2\pi$ | $+0.001\ 161\ 409\ 733$ |
| magnetic | $+\alpha^2/16\pi^2$ | $+0.000\ 000\ 337\ 218$ |
| weak$_1$ | $+\alpha^2/4\pi$ | $+0.000\ 004\ 237\ 608$ |
| weak$_2$ | $-10\alpha^3/4\pi$ | $-0.000\ 000\ 309\ 233$ |
| weak$_3$ | $+10(32\alpha)^1\,\alpha^3/4\pi$ | $+0.000\ 000\ 072\ 210$ |
| weak$_4$ | $-10(32\alpha)^3\,\alpha^3/4\pi$ | $-0.000\ 000\ 005\ 937$ |
| weak$_5$ | $+10(32\alpha)^4\,\alpha^3/4\pi$ | $+0.000\ 000\ 001\ 919$ |
| weak$_6$ | $-10(32\alpha)^5\,\alpha^3/4\pi$ | $-0.000\ 000\ 000\ 214$ |
| weak$_7$ | $+10(32\alpha)^6\,\alpha^3/4\pi$ | $+0.000\ 000\ 000\ 050$ |
| weak$_8$ | $-10(32\alpha)^7\,\alpha^3/4\pi$ | $-0.000\ 000\ 000\ 011$ |
| weak$_9$ | $+10(32\alpha)^8\,\alpha^3/4\pi$ | $+0.000\ 000\ 000\ 002$ |
| weak$_{10}$ | $-10(32\alpha)^9\,\alpha^3/4\pi$ | $-0.000\ 000\ 000\ 000$ |
| weak$_{11}$ | $+10(32\alpha)^{10}\,\alpha^3/4\pi$ | $+0.000\ 000\ 000\ 000$ |

total calculated g/2 factor for –e$_3$     $+0.588\ 510\ 166\ 228$

Table 5-10
Positron family
-e$_4$ g/2 Factor Evaluation with 2010 α

| force: | g/2 factor term: | numerical value: |
|---|---|---|
| strong | $+1$ | $+1.000\ 000\ 000\ 000$ |
| meso-electric | $-bn\pi\alpha$ | $-0.917\ 012\ 368\ 931$ |
| electric | $+\alpha/2\pi$ | $+0.001\ 161\ 409\ 733$ |
| magnetic | $+\alpha^2/16\pi^2$ | $+0.000\ 000\ 337\ 218$ |
| weak$_1$ | $-\alpha^2/4\pi$ | $-0.000\ 004\ 237\ 608$ |
| weak$_2$ | $+15\alpha^3/4\pi$ | $+0.000\ 000\ 463\ 849$ |
| weak$_3$ | $-15(32\alpha)^1\alpha^3/4\pi$ | $-0.000\ 000\ 103\ 316$ |
| weak$_4$ | $+15(32\alpha)^3\alpha^3/4\pi$ | $+0.000\ 000\ 005\ 633$ |
| weak$_5$ | $-15(32\alpha)^4\alpha^3/4\pi$ | $-0.000\ 000\ 001\ 315$ |
| weak$_6$ | $+15(32\alpha)^5\alpha^3/4\pi$ | $+0.000\ 000\ 000\ 307$ |
| weak$_7$ | $-15(32\alpha)^6\alpha^3/4\pi$ | $-0.000\ 000\ 000\ 071$ |
| weak$_8$ | $+15(32\alpha)^7\alpha^3/4\pi$ | $+0.000\ 000\ 000\ 016$ |
| weak$_9$ | $-15(32\alpha)^8\alpha^3/4\pi$ | $-0.000\ 000\ 000\ 003$ |
| weak$_{10}$ | $+15(32\alpha)^9\alpha^3/4\pi$ | $+0.000\ 000\ 000\ 000$ |
| weak$_{11}$ | $-15(32\alpha)^{10}\alpha^3/4\pi$ | $-0.000\ 000\ 000\ 000$ |

total calculated g/2 factor for −e$_4$     $+0.084\ 145\ 505\ 542$

Table 5-11
Pion family
Pion $\pi_1$ g/2 Factor Evaluation with 2010 $\alpha$

| force: | g/2 factor term: | numerical value: |
|---|---|---|
| strong | $+1$ | $+1.000\ 000\ 000\ 000$ |
| meso-electric | $-(b-1)n\pi\alpha$ | $-0.000\ 000\ 000\ 000$ |
| electric | $+\alpha/2\pi$ | $+0.001\ 161\ 409\ 733$ |
| magnetic | $+\alpha^2/16\pi^2$ | $+0.000\ 000\ 337\ 218$ |
| weak$_1$ | $-\alpha^2/4\pi$ | $-0.000\ 004\ 237\ 608$ |
| weak$_2$ | $+\alpha^3/4\pi$ | $+0.000\ 000\ 030\ 923$ |
| weak$_3$ | $-(32\alpha)^1\,\alpha^3/4\pi$ | $-0.000\ 000\ 007\ 221$ |
| weak$_4$ | $+(32\alpha)^3\,\alpha^3/4\pi$ | $+0.000\ 000\ 000\ 393$ |
| weak$_5$ | $-(32\alpha)^4\,\alpha^3/4\pi$ | $-0.000\ 000\ 000\ 091$ |
| weak$_6$ | $+(32\alpha)^5\,\alpha^3/4\pi$ | $+0.000\ 000\ 000\ 021$ |
| weak$_7$ | $-(32\alpha)^6\,\alpha^3/4\pi$ | $-0.000\ 000\ 000\ 005$ |
| weak$_8$ | $+(32\alpha)^7\,\alpha^3/4\pi$ | $+0.000\ 000\ 000\ 001$ |
| weak$_9$ | $-(32\alpha)^8\,\alpha^3/4\pi$ | $-0.000\ 000\ 000\ 000$ |

| total calculated g/2 factor for pion | $+1.001\ 157\ 533\ 365$ |
|---|---|

Table 5-12
Pion family
Kaon $\pi_2$ g/2 Factor Evaluation with 2010 $\alpha$

| force: | g/2 factor term: | numerical value: |
|---|---|---|
| strong | $+1$ | $+1.000\ 000\ 000\ 000$ |
| meso-electric | $-(b-1)n\pi\alpha$ | $-0.022\ 925\ 309\ 223$ |
| electric | $+\alpha/2\pi$ | $+0.001\ 161\ 409\ 733$ |
| magnetic | $+\alpha^2/16\pi^2$ | $+0.000\ 000\ 337\ 218$ |
| $weak_1$ | $+\alpha^2/4\pi$ | $+0.000\ 004\ 237\ 608$ |
| $weak_2$ | $-3\alpha^3/4\pi$ | $-0.000\ 000\ 092\ 769$ |
| $weak_3$ | $+3(32\alpha)^1\alpha^3/4\pi$ | $+0.000\ 000\ 021\ 663$ |
| $weak_4$ | $-3(32\alpha)^3\alpha^3/4\pi$ | $-0.000\ 000\ 001\ 181$ |
| $weak_5$ | $+3(32\alpha)^4\alpha^3/4\pi$ | $+0.000\ 000\ 000\ 275$ |
| $weak_6$ | $-3(32\alpha)^5\alpha^3/4\pi$ | $-0.000\ 000\ 000\ 064$ |
| $weak_7$ | $+3(32\alpha)^6\alpha^3/4\pi$ | $+0.000\ 000\ 000\ 015$ |
| $weak_8$ | $-3(32\alpha)^7\alpha^3/4\pi$ | $-0.000\ 000\ 000\ 003$ |
| $weak_9$ | $+3(32\alpha)^8\alpha^3/4\pi$ | $+0.000\ 000\ 000\ 000$ |
| $weak_{10}$ | $-3(32\alpha)^9\alpha^3/4\pi$ | $-0.000\ 000\ 000\ 000$ |

total calculated g/2 factor for $\pi_2$     $+0.978\ 240\ 603\ 272$

Table 5-13
Pion family
D-on $\pi_3$ g/2 Factor Evaluation with 2010 $\alpha$

| force: | g/2 factor term: | numerical value: |
|---|---|---|
| strong | $+1$ | $+1.000\ 000\ 000\ 000$ |
| meso-electric | $-(b-1)n\pi\alpha$ | $-0.137\ 551\ 874\ 189$ |
| electric | $+\alpha/2\pi$ | $+0.001\ 161\ 409\ 727$ |
| magnetic | $+\alpha^2/16\pi^2$ | $+0.000\ 000\ 337\ 218$ |
| $weak_1$ | $-\alpha^2/4\pi$ | $-0..000\ 004\ 237\ 608$ |
| $weak_2$ | $+6\alpha^3/4\pi$ | $+0.000\ 000\ 185\ 539$ |
| $weak_3$ | $-6(32\alpha)^1\,\alpha^3/4\pi$ | $-0.000\ 000\ 043\ 663$ |
| $weak_4$ | $+6(32\alpha)^3\,\alpha^3/4\pi$ | $+0.000\ 000\ 002\ 362$ |
| $weak_5$ | $-6(32\alpha)^4\,\alpha^3/4\pi$ | $-0.000\ 000\ 000\ 551$ |
| $weak_6$ | $+6(32\alpha)^5\,\alpha^3/4\pi$ | $+0.000\ 000\ 000\ 128$ |
| $weak_7$ | $-6(32\alpha)^6\,\alpha^3/4\pi$ | $-0.000\ 000\ 000\ 030$ |
| $weak_8$ | $+6(32\alpha)^7\,\alpha^3/4\pi$ | $+0.000\ 000\ 000\ 007$ |
| $weak_9$ | $-6(32\alpha)^8\,\alpha^3/4\pi$ | $-0.000\ 000\ 000\ 001$ |
| $weak_{10}$ | $+6(32\alpha)^9\,\alpha^3/4\pi$ | $+0.000\ 000\ 000\ 000$ |

total calculated g/2 factor for $\pi_3$  $+0.863\ 605\ 778\ 946$

Table 5-14
Pion family
$\pi_4$ g/2 Factor Evaluation with 2010 $\alpha$

| force: | g/2 factor term: | numerical value: |
|---|---|---|
| strong | $+1$ | $+1.000\ 000\ 000\ 000$ |
| meso-electric | $-(b-1)n\pi\alpha$ | $-0.412\ 655\ 566\ 198$ |
| electric | $+\alpha/2\pi$ | $+0.001\ 161\ 409\ 733$ |
| magnetic | $+\alpha^2/16\pi^2$ | $+0.000\ 000\ 337\ 218$ |
| $weak_1$ | $+\alpha^2/4\pi$ | $+0.000\ 004\ 237\ 608$ |
| $weak_2$ | $-10\alpha^3/4\pi$ | $-0.000\ 000\ 309\ 233$ |
| $weak_3$ | $+10(32\alpha)^1\alpha^3/4\pi$ | $+0.000\ 000\ 072\ 210$ |
| $weak_4$ | $-10(32\alpha)^3\alpha^3/4\pi$ | $-0.000\ 000\ 005\ 937$ |
| $weak_5$ | $+10(32\alpha)^4\alpha^3/4\pi$ | $+0.000\ 000\ 001\ 919$ |
| $weak_6$ | $-10(32\alpha)^5\alpha^3/4\pi$ | $-0.000\ 000\ 000\ 214$ |
| $weak_7$ | $+10(32\alpha)^6\alpha^3/4\pi$ | $+0.000\ 000\ 000\ 050$ |
| $weak_8$ | $-10(32\alpha)^7\alpha^3/4\pi$ | $-0.000\ 000\ 000\ 011$ |
| $weak_9$ | $+10(32\alpha)^8\alpha^3/4\pi$ | $+0.000\ 000\ 000\ 002$ |
| $weak_{10}$ | $-10(32\alpha)^9\alpha^3/4\pi$ | $-0.000\ 000\ 000\ 000$ |
| $weak_{11}$ | $+10(32\alpha)^{10}\alpha^3/4\pi$ | $+0.000\ 000\ 000\ 000$ |

total calculated g/2 factor for $\pi_4$     $+0.588\ 510\ 179\ 146$

Table 5-15
Pion family
$\pi_5$ g/2 Factor Evaluation with 2010 $\alpha$

| force: | g/2 factor term: | numerical value: |
|---|---|---|
| strong | $+1$ | +1.000 000 000 000 |
| meso-electric | $-(b-1)n\pi\alpha$ | -0.917 012 368 931 |
| electric | $+\alpha/2\pi$ | +0.001 161 409 733 |
| magnetic | $+\alpha^2/16\pi^2$ | +0.000 000 337 218 |
| $weak_1$ | $-\alpha^2/4\pi$ | -0.000 004 237 608 |
| $weak_2$ | $+15\alpha^3/4\pi$ | +0.000 000 463 849 |
| $weak_3$ | $-15(32\alpha)^1\alpha^3/4\pi$ | -0.000 000 103 316 |
| $weak_4$ | $+15(32\alpha)^3\alpha^3/4\pi$ | +0.000 000 005 633 |
| $weak_5$ | $-15(32\alpha)^4\alpha^3/4\pi$ | -0.000 000 001 315 |
| $weak_6$ | $+15(32\alpha)^5\alpha^3/4\pi$ | +0.000 000 000 307 |
| $weak_7$ | $-15(32\alpha)^6\alpha^3/4\pi$ | -0.000 000 000 071 |
| $weak_8$ | $+15(32\alpha)^7\alpha^3/4\pi$ | +0.000 000 000 016 |
| $weak_9$ | $-15(32\alpha)^8\alpha^3/4\pi$ | -0.000 000 000 003 |
| $weak_{10}$ | $+15(32\alpha)^9\alpha^3/4\pi$ | +0.000 000 000 000 |
| $weak_{11}$ | $-15(32\alpha)^{10}\alpha^3/4\pi$ | -0.000 000 000 000 |

total calculated g/2 factor for $\pi_5$     +0.084 145 505 512

Table 5-16
Anti-pion family
Anti-pion $-\pi_1$ g/2 Factor Evaluation with 20010 $\alpha$

| force: | g/2 factor term: | numerical value: |
|---|---|---|
| strong | $-1$ | -1.000 000 000 000 |
| electric | $-\alpha/2\pi$ | -0.001 161 409 733 |
| magnetic | $-\alpha^2/16\pi^2$ | -0.000 000 337 218 |
| $weak_1$ | $+\alpha^2/4\pi$ | +0.000 004 237 608 |
| $weak_2$ | $-\alpha^3/4\pi$ | -0.000 000 030 923 |
| $weak_3$ | $+(32\alpha)^1\,\alpha^3/4\pi$ | +0.000 000 007 221 |
| $weak_4$ | $-(32\alpha)^3\,\alpha^3/4\pi$ | -0.000 000 000 393 |
| $weak_5$ | $+(32\alpha)^4\,\alpha^3/4\pi$ | +0.000 000 000 091 |
| $weak_6$ | $-(32\alpha)^5\,\alpha^3/4\pi$ | -0.000 000 000 021 |
| $weak_7$ | $+(32\alpha)^6\,\alpha^3/4\pi$ | +0.000 000 000 005 |
| $weak_8$ | $-(32\alpha)^7\,\alpha^3/4\pi$ | -0.000 000 000 001 |
| $weak_9$ | $+(32\alpha)^8\,\alpha^3/4\pi$ | +0.000 000 000 000 |

total calculated g/2 factor for $-\pi_1$     -1.001 157 533 365

Table 5-17
Anti-pion family
Anti-kaon $-\pi_2$ g/2 Factor Evaluation with 2010 $\alpha$

| force: | g/2 factor term: | numerical value: |
|---|---|---|
| strong | $-1$ | -1.000 000 000 000 |
| electric | $-\alpha/2\pi$ | -0.001 161 409 733 |
| magnetic | $-\alpha^2/16\pi^2$ | -0.000 000 337 218 |
| $weak_1$ | $-\alpha^2/4\pi$ | -0.000 004 237 608 |
| $weak_2$ | $+3\alpha^3/4\pi$ | +0.000 000 092 769 |
| $weak_3$ | $-3(32\alpha)^1\alpha^3/4\pi$ | -0.000 000 021 663 |
| $weak_4$ | $+3(32\alpha)^3\alpha^3/4\pi$ | +0.000 000 001 181 |
| $weak_5$ | $-3(32\alpha)^4\alpha^3/4\pi$ | -0.000 000 000 275 |
| $weak_6$ | $+3(32\alpha)^5\alpha^3/4\pi$ | +0.000 000 000 064 |
| $weak_7$ | $-3(32\alpha)^6\alpha^3/4\pi$ | -0.000 000 000 015 |
| $weak_8$ | $+3(32\alpha)^7\alpha^3/4\pi$ | +0.000 000 000 003 |
| $weak_9$ | $-3(32\alpha)^8\alpha^3/4\pi$ | -0.000 000 000 000 |
| $weak_{10}$ | $+3(32\alpha)^9\alpha^3/4\pi$ | +0.000 000 000 000 |

total calculated g/2 factor for $-\pi_2$    -1.001 165 912 495

Table 5-18
Anti-pion family
Anti-D-on $-\pi_3$ g/2 Factor Evaluation with 2010 $\alpha$

| force: | g/2 factor term: | numerical value: |
|---|---|---|
| strong | $-1$ | -1.000 000 000 000 |
| electric | $-\alpha/2\pi$ | -0.001 161 409 733 |
| magnetic | $-\alpha^2/16\pi^2$ | -0.000 000 337 218 |
| $weak_1$ | $+\alpha^2/4\pi$ | +0.000 004 237 608 |
| $weak_2$ | $-6\alpha^3/4\pi$ | -0.000 000 185 539 |
| $weak_3$ | $+6(32\alpha)^1\alpha^3/4\pi$ | +0.000 000 043 663 |
| $weak_4$ | $-6(32\alpha)^3\alpha^3/4\pi$ | -0.000 000 002 362 |
| $weak_5$ | $+6(32\alpha)^4\alpha^3/4\pi$ | +0.000 000 000 551 |
| $weak_6$ | $-6(32\alpha)^5\alpha^3/4\pi$ | -0.000 000 000 128 |
| $weak_7$ | $+6(32\alpha)^6\alpha^3/4\pi$ | +0.000 000 000 030 |
| $weak_8$ | $-6(32\alpha)^7\alpha^3/4\pi$ | -0.000 000 000 007 |
| $weak_9$ | $+6(32\alpha)^8\alpha^3/4\pi$ | +0.000 000 000 001 |
| $weak_{10}$ | $-6(32\alpha)^7\alpha^3/4\pi$ | -0.000 000 000 000 |

total calculated g/2 factor for $-\pi_3$    -1.001 157 653 136

Table 5-19
Anti-pion family
$-\pi_4$ g/2 Factor Evaluation with 2010 $\alpha$

| force: | g/2 factor term: | numerical value: |
|---|---|---|
| strong | $-1$ | -1.000 000 000 000 |
| electric | $-\alpha/2\pi$ | -0.001 161 409 733 |
| magnetic | $-\alpha^2/16\pi^2$ | -0.000 000 337 218 |
| $weak_1$ | $-\alpha^2/4\pi$ | -0.000 004 237 608 |
| $weak_2$ | $+10\alpha^3/4\pi$ | +0.000 000 309 233 |
| $weak_3$ | $-10(32\alpha)^1\alpha^3/4\pi$ | -0.000 000 072 210 |
| $weak_4$ | $+10(32\alpha)^3\alpha^3/4\pi$ | +0.000 000 005 937 |
| $weak_5$ | $-10(32\alpha)^4\alpha^3/4\pi$ | -0.000 000 001 919 |
| $weak_6$ | $+10(32\alpha)^5\alpha^3/4\pi$ | +0.000 000 000 214 |
| $weak_7$ | $-10(32\alpha)^6\alpha^3/4\pi$ | -0.000 000 000 050 |
| $weak_8$ | $+10(32\alpha)^7\alpha^3/4\pi$ | +0.000 000 000 011 |
| $weak_9$ | $-10(32\alpha)^8\alpha^3/4\pi$ | -0.000 000 000 002 |
| $weak_{10}$ | $+10(32\alpha)^9\alpha^3/4\pi$ | +0.000 000 000 000 |
| $weak_{11}$ | $-10(32\alpha)^{10}\alpha^3/4\pi$ | -0.000 000 000 000 |

total calculated g/2 factor for $-\pi_4$     -1.001 165 744 345

Table 5-20
Anti-pion family
$-\pi_5$ g/2 Factor Evaluation with 20010 $\alpha$

| force: | g/2 factor term: | numerical value: |
|---|---|---|
| strong | -1 | -1.000 000 000 000 |
| electric | $-\alpha/2\pi$ | -0.001 161 409 733 |
| magnetic | $-\alpha^2/16\pi^2$ | -0.000 000 337 218 |
| weak$_1$ | $+\alpha^2/4\pi$ | + 0.000 004 237 608 |
| weak$_2$ | $-15\alpha^3/4\pi$ | -0.000 000 463 849 |
| weak$_3$ | $+15(32\alpha)^1\alpha^3/4\pi$ | +0.000 000 103 316 |
| weak$_4$ | $-15(32\alpha)^3\alpha^3/4\pi$ | -0.000 000 005 633 |
| weak$_5$ | $+15(32\alpha)^4\alpha^3/4\pi$ | +0.000 000 001 315 |
| weak$_6$ | $-15(32\alpha)^5\alpha^3/4\pi$ | -0.000 000 000 307 |
| weak$_7$ | $+15(32\alpha)^6\alpha^3/4\pi$ | +0.000 000 000 071 |
| weak$_8$ | $-15(32\alpha)^7\alpha^3/4\pi$ | -0.000 000 000 016 |
| weak$_9$ | $+15(32\alpha)^8\alpha^3/4\pi$ | +0.000 000 000 003 |
| weak$_{10}$ | $-15(32\alpha)^9\alpha^3/4\pi$ | -0.000 000 000 000 |
| weak$_{11}$ | $+15(32\alpha)^{10}\alpha^3/4\pi$ | +0.000 000 000 000 |

total calculated g/2 factor for $-\pi_5$    -1.001 157 869 444

Table 5-21
Neutron family
Neutron $n_1$ g/2 Factor Evaluation with 2010 $\alpha$

| force: | g/2 factor term: | numerical value: |
|---|---|---|
| strong | -1 | -1.000 000 000 000 |
| meso-electric | $-bn\pi\alpha$ | -0.022 925 309 222 |
| electric | $-\alpha/2\pi$ | -0.001 161 409 733 |
| magnetic | $-\alpha^2/16\pi^2$ | -0.000 000 337 218 |
| $weak_1$ | $+\alpha^2/8\pi$ | +0.000 002 118 804 |
| $weak_2$ | $-3\alpha^3/4\pi$ | -0.000 000 092 770 |
| $weak_3$ | $+3(32\alpha)^1\alpha^3/4\pi$ | +0.000 000 021 831 |
| $weak_4$ | $-3(32\alpha)^3\alpha^3/4\pi$ | -0.000 000 001 181 |
| $weak_5$ | $+3(32\alpha)^4\alpha^3/4\pi$ | +0.000 000 000 275 |
| $weak_6$ | $-3(32\alpha)^5\alpha^3/4\pi$ | -0.000 000 000 064 |
| $weak_7$ | $+3(32\alpha)^6\alpha^3/4\pi$ | +0.000 000 000 015 |
| $weak_8$ | $-3(32\alpha)^7\alpha^3/4\pi$ | -0.000 000 000 003 |
| $weak_9$ | $+3(32\alpha)^8\alpha^3/4\pi$ | +0.000 000 000 000 |
| $weak_{10}$ | $-3(32\alpha)^9\alpha^3/4\pi$ | -0.000 000 000 000 |

total calculated g/2 factor for neutron  -1.024 085 009 266

Table 5-22
Neutron family
$\Lambda$   $n_2$ g/2 Factor Evaluation with 2010 $\alpha$

| force: | g/2 factor term: | numerical value: |
|---|---|---|
| strong | $-1$ | -1.000 000 000 000 |
| meso-electric | $-bn\pi\alpha$ | -0.137 551 855 339 |
| electric | $-\alpha/2\pi$ | -0.001 161 409 733 |
| magnetic | $-\alpha^2/16\pi^2$ | -0.000 000 337 218 |
| $weak_1$ | $-\alpha^2/4\pi$ | -0.000 004 237 608 |
| $weak_2$ | $+6\alpha^3/4\pi$ | +0.000 000 185 539 |
| $weak_3$ | $-6(32\alpha)^1\alpha^3/4\pi$ | -0.000 000 043 663 |
| $weak_4$ | $+6(32\alpha)^3\alpha^3/4\pi$ | +0.000 000 002 362 |
| $weak_5$ | $-6(32\alpha)^4\alpha^3/4\pi$ | -0.000 000 000 551 |
| $weak_6$ | $+6(32\alpha)^5\alpha^3/4\pi$ | +0.000 000 000 128 |
| $weak_7$ | $-6(32\alpha)^6\alpha^3/4\pi$ | -0.000 000 000 030 |
| $weak_8$ | $+6(32\alpha)^7\alpha^3/4\pi$ | +0.000 000 000 007 |
| $weak_9$ | $-6(32\alpha)^8\alpha^3/4\pi$ | -0.000 000 000 001 |
| $weak_{10}$ | $+6(32\alpha)^9\alpha^3/4\pi$ | +0.000 000 000 000 |

total calculated g/2 factor for $n_2$      -1.138 716 794 286

Table 5-23
Neutron family
$\Sigma^0$ $n_3$ $_g/2$ Factor Evaluation with 2010 $\alpha$

| force: | g/2 factor term: | numerical value: |
|---|---|---|
| strong | $-1$ | -1.000 000 000 000 |
| meso-electric | $-bn\pi\alpha$ | -0.412 655 565 996 |
| electric | $-\alpha/2\pi$ | -0.001 161 409 733 |
| magnetic | $-\alpha^2/16\pi^2$ | -0.000 000 337 218 |
| $weak_1$ | $+\alpha^2/4\pi$ | +0.000 004 237 608 |
| $weak_2$ | $-10\alpha^3/4\pi$ | -0.000 000 309 233 |
| $weak_3$ | $+10(32\alpha)^1\alpha^3/4\pi$ | +0.000 000 072 210 |
| $weak_4$ | $-10(32\alpha)^3\alpha^3/4\pi$ | -0.000 000 005 937 |
| $weak_5$ | $+10(32\alpha)^4\alpha^3/4\pi$ | +0.000 000 001 919 |
| $weak_6$ | $-10(32\alpha)^5\alpha^3/4\pi$ | -0.000 000 000 214 |
| $weak_7$ | $+10(32\alpha)^6\alpha^3/4\pi$ | +0.000 000 000 050 |
| $weak_8$ | $-10(32\alpha)^7\alpha^3/4\pi$ | -0.000 000 000 011 |
| $weak_9$ | $+10(32\alpha)^8\alpha^3/4\pi$ | +0.000 000 000 002 |
| $weak_{10}$ | $-10(32\alpha)^9\alpha^3/4\pi$ | -0.000 000 000 000 |
| $weak_{11}$ | $+10(32\alpha)^{10}\alpha^3/4\pi$ | +0.000 000 000 000 |

total calculated g/2 factor for $n_3$     -1.413 813 308 461

Table 5-24
Neutron family
$\Lambda_b^0$   n4 g/2 Factor Evaluation with 2010 α

| force: | g/2 factor term: | numerical value: |
|---|---|---|
| strong | $-1$ | -1.000 000 000 000 |
| meso-electric | $-bn\pi\alpha$ | -0.917 012 368 931 |
| electric | $-\alpha/2\pi$ | -0.001 161 409 733 |
| magnetic | $-\alpha^2/16\pi^2$ | -0.000 000 337 218 |
| $weak_1$ | $-\alpha^2/4\pi$ | -0.000 004 237 608 |
| $weak_2$ | $+15\alpha^3/4\pi$ | +0.000 000 463 849 |
| $weak_3$ | $-15(32\alpha)^1\alpha^3/4\pi$ | -0.000 000 103 316 |
| $weak_4$ | $+15(32\alpha)^3\alpha^3/4\pi$ | +0.000 000 005 633 |
| $weak_5$ | $-15(32\alpha)^4\alpha^3/4\pi$ | -0.000 000 001 315 |
| $weak_6$ | $+15(32\alpha)^5\alpha^3/4\pi$ | +0.000 000 000 307 |
| $weak_7$ | $-15(32\alpha)^6\alpha^3/4\pi$ | -0.000 000 000 071 |
| $weak_8$ | $+15(32\alpha)^7\alpha^3/4\pi$ | +0.000 000 000 016 |
| $weak_9$ | $-15(32\alpha)^8\alpha^3/4\pi$ | -0.000 000 000 003 |
| $weak_{10}$ | $+15(32\alpha)^9\alpha^3/4\pi$ | +0.000 000 000 000 |
| $weak_{11}$ | $-15(32\alpha)^{10}\alpha^3/4\pi$ | -0.000 000 000 000 |

total calculated g/2 factor for n4      -1.918 178 088 390

Table 5-25
Anti-neutron family
Anti-neutron $-n_1$ g/2 Factor Evaluation with 2010 $\alpha$

| force: | g/2 factor term: | numerical value: |
|---|---|---|
| strong | $+1$ | $+1.000\ 000\ 000\ 000$ |
| electric | $+\alpha / 2\pi$ | $+0.001\ 161\ 409\ 733$ |
| magnetic | $+\alpha^2 / 16\pi^2$ | $+0.000\ 000\ 337\ 218$ |
| $weak_1$ | $-\alpha^2 / 4\pi$ | $-0.000\ 004\ 237\ 608$ |
| $weak_2$ | $+3\alpha^3 / 4\pi$ | $+0.000\ 000\ 092\ 769$ |
| $weak_3$ | $-3(32\alpha)^1 \alpha^3 / 4\pi$ | $-0.000\ 000\ 021\ 663$ |
| $weak_4$ | $+3(32\alpha)^3 \alpha^3 / 4\pi$ | $+0.000\ 000\ 001\ 181$ |
| $weak_5$ | $-3(32\alpha)^4 \alpha^3 / 4\pi$ | $-0.000\ 000\ 000\ 275$ |
| $weak_6$ | $+3(32\alpha)^5 \alpha^3 / 4\pi$ | $+0.000\ 000\ 000\ 064$ |
| $weak_7$ | $-3(32\alpha)^6 \alpha^3 / 4\pi$ | $-0.000\ 000\ 000\ 015$ |
| $weak_8$ | $+3(32\alpha)^7 \alpha^3 / 4\pi$ | $+0.000\ 000\ 000\ 003$ |
| $weak_9$ | $-3(32\alpha)^8 \alpha^3 / 4\pi$ | $-0.000\ 000\ 000\ 000$ |
| $weak_{10}$ | $+3(32\alpha)^9 \alpha^3 / 4\pi$ | $+0.000\ 000\ 000\ 000$ |

total calculated g/2 factor for $-n_1$    $+1.001\ 157\ 581\ 408$

Table 5-26
Anti-neutron family
-Λ -n$_2$ g/2 Factor Evaluation with 2010 α

| force: | g/2 factor term: | numerical value: |
|---|---|---|
| strong | $+1$ | $+1.000\ 000\ 000\ 000$ |
| electric | $+\alpha/2\pi$ | $+0.001\ 161\ 409\ 733$ |
| magnetic | $+\alpha^2/16\pi^2$ | $+0.000\ 000\ 337\ 218$ |
| weak$_1$ | $+\alpha^2/4\pi$ | $+0.000\ 004\ 237\ 608$ |
| weak$_2$ | $-6\alpha^3/4\pi$ | $-0.000\ 000\ 185\ 539$ |
| weak$_3$ | $+6(32\alpha)^1\,\alpha^3/4\pi$ | $+0.000\ 000\ 043\ 663$ |
| weak$_4$ | $-6(32\alpha)^3\,\alpha^3/4\pi$ | $-0.000\ 000\ 002\ 362$ |
| weak$_5$ | $+6(32\alpha)^4\,\alpha^3/4\pi$ | $+0.000\ 000\ 000\ 551$ |
| weak$_6$ | $-6(32\alpha)^5\,\alpha^3/4\pi$ | $-0.000\ 000\ 000\ 128$ |
| weak$_7$ | $+6(32\alpha)^6\,\alpha^3/4\pi$ | $+0.000\ 000\ 000\ 030$ |
| weak$_8$ | $-6(32\alpha)^7\,\alpha^3/4\pi$ | $-0.000\ 000\ 000\ 007$ |
| weak$_9$ | $+6(32\alpha)^8\,\alpha^3/4\pi$ | $+0.000\ 000\ 000\ 001$ |
| weak$_{10}$ | $-6(32\alpha)^7\,\alpha^3/4\pi$ | $-0.000\ 000\ 000\ 000$ |

total calculated g/2 factor for $-$n$_2$     $+1.001\ 165\ 840\ 767$

Table 5-27
Anti-neutron family
$-\Sigma^0$ -n$_3$   g/2 Factor Evaluation with 2010 $\alpha$

| force: | g/2 factor term: | numerical value: |
|---|---|---|
| strong | $+1$ | $+1.000\ 000\ 000\ 000$ |
| electric | $+\alpha/2\pi$ | $+0.001\ 161\ 409\ 733$ |
| magnetic | $+\alpha^2/16\pi^2$ | $+0.000\ 000\ 337\ 218$ |
| weak$_1$ | $-\alpha^2/4\pi$ | $-0.000\ 004\ 237\ 608$ |
| weak$_2$ | $+10\alpha^3/4\pi$ | $+0.000\ 000\ 309\ 233$ |
| weak$_3$ | $-10(32\alpha)^1\,\alpha^3/4\pi$ | $-0.000\ 000\ 072\ 210$ |
| weak$_4$ | $+10(32\alpha)^3\,\alpha^3/4\pi$ | $+0.000\ 000\ 005\ 937$ |
| weak$_5$ | $-10(32\alpha)^4\,\alpha^3/4\pi$ | $-0.000\ 000\ 001\ 919$ |
| weak$_6$ | $+10(32\alpha)^5\,\alpha^3/4\pi$ | $+0.000\ 000\ 000\ 214$ |
| weak$_7$ | $-10(32\alpha)^6\,\alpha^3/4\pi$ | $-0.000\ 000\ 000\ 050$ |
| weak$_8$ | $+10(32\alpha)^7\,\alpha^3/4\pi$ | $+0.000\ 000\ 000\ 011$ |
| weak$_9$ | $-10(32\alpha)^8\,\alpha^3/4\pi$ | $-0.000\ 000\ 000\ 002$ |
| weak$_{10}$ | $+10(32\alpha)^9\,\alpha^3/4\pi$ | $+0.000\ 000\ 000\ 000$ |
| weak$_{11}$ | $-10(32\alpha)^{10}\,\alpha^3/4\pi$ | $-0.000\ 000\ 000\ 000$ |

total calculated g/2 factor for $-$n$_3$   $+1.001\ 157\ 750\ 558$

Table 5-28
Anti-neutron family
$-\Lambda_b^0$ $-n_4$  g/2 Factor Evaluation with 2010 $\alpha$

| force: | g/2 factor term: | numerical value: |
|---|---|---|
| strong | $+1$ | $+1.000\ 000\ 000\ 000$ |
| electric | $+\alpha/2\pi$ | $+0.001\ 161\ 409\ 733$ |
| magnetic | $+\alpha^2/16\pi^2$ | $+0.000\ 000\ 337\ 218$ |
| $weak_1$ | $+\alpha^2/4\pi$ | $+\ 0.000\ 004\ 237\ 608$ |
| $weak_2$ | $-15\alpha^3/4\pi$ | $-0.000\ 000\ 463\ 849$ |
| $weak_3$ | $+15(32\alpha)^1\alpha^3/4\pi$ | $+0.000\ 000\ 103\ 316$ |
| $weak_4$ | $-15(32\alpha)^3\alpha^3/4\pi$ | $-0.000\ 000\ 005\ 633$ |
| $weak_5$ | $+15(32\alpha)^4\alpha^3/4\pi$ | $+0.000\ 000\ 001\ 315$ |
| $weak_6$ | $-15(32\alpha)^5\alpha^3/4\pi$ | $-0.000\ 000\ 000\ 307$ |
| $weak_7$ | $+15(32\alpha)^6\alpha^3/4\pi$ | $+0.000\ 000\ 000\ 071$ |
| $weak_8$ | $-15(32\alpha)^7\alpha^3/4\pi$ | $-0.000\ 000\ 000\ 016$ |
| $weak_9$ | $+15(32\alpha)^8\alpha^3/4\pi$ | $+0.000\ 000\ 000\ 003$ |
| $weak_{10}$ | $-15(32\alpha)^9\alpha^3/4\pi$ | $-0.000\ 000\ 000\ 000$ |
| $weak_{11}$ | $+15(32\alpha)^{10}\alpha^3/4\pi$ | $+0.000\ 000\ 000\ 000$ |

total calculated g/2 factor for $-n_4$    $+1.001\ 165\ 619\ 459$

www.ingramcontent.com/pod-product-compliance
Lightning Source LLC
Chambersburg PA
CBHW070931180526
45168CB00003B/1025